THE MONT BLANC RANGE
Classic Snow, Ice and Mixed Climbs

Note for the UK / US Edition: This guidebook describes thirty-six classic, lower grade climbs in the Mont Blanc Range – the busiest, highest and in many ways the most challenging of all the western alpine ranges. The climbs have all been climbed or checked recently by the authors in the aftermath of the very hot summer of 2003 when major topographical changes took place in the range. The climbs are grade F (*Facile*) to AD (*Assez Difficile*) with three climbs of D (*Difficile*). Though of easy to middling *technical* difficulty it should be noted that all of these climbs are on high and complex alpine peaks that can become very difficult and serious in the event of poor weather or deteriorating snow and ice conditions. All require a high level of alpine competence with nighttime glacier approaches, use of crampons and ice axes, belaying and self arrest skills, abseiling proficiency and crevasse rescue knowledge. The climbs are listed below in their various grades (with their last recorded Alpine Club grades in brackets). A good, well-compacted snow cover tends to make the climbs easier.

Cover photo: Dent du Géant. All photographs in this work, except those on the back cover, are by Jean-Louis Laroche and Florence LeLong. All diagrams and watercolours are by Florence LeLong (except the general map on page 5). Although a few English words have been added, the map/diagram annotations remain mainly in French to match the proper names on local maps and also with *rimayes* and *rappels* for the *bergschrunds* and *abseils* of the main text.

Thanks to François Damilano, Françoise Rouxel, Guillaume Bernole (PGHM), Pierre and Evelyne Allain, Monique, Patricia and Sylvia (OHM)

First published in 2006 by Editions Glenat, Grenoble   © 2006 Éditions Glénat

First English language edition published jointly in Great Britain and the U.S.A. in 2007
by Bâton Wicks Publications, London and
Menasha Ridge Press, Birmingham, Alabama

All trade enquiries in the UK, Europe and Commonwealth (except Canada) to:
Cordee, 3a deMontfort Street, Leicester, LE1 7HD, UK
All trade enquiries in the USA and Canada to:
Publishers Group West; all other orders may be directed to www.menasharidge.com

British Library and Cataloguing in Publication Data:
A catalogue record for this book exists at the British Library ISBN 1-898573-72-7 / 978-1-898573-72-2
Library of Congress Cataloging in Publication Data
A catalog record of this book exists at the Library of Congress ISBN 0-89732-651-2 / 978-0-89732-651-3

Jean-Louis Laroche — Florence Lelong

THE MONT BLANC RANGE
Classic Snow, Ice and Mixed Climbs

Adapted from a translation by Blyth Wright

Bâton Wicks
London U.K.

Menasha Ridge Press
Birmingham Alabama U.S.A.

✳ CONTENTS

See also detailed sketches for each route description

col de Balme

Vallorcine
Martigny
Suisse

col des
Montets

LE TOUR

cabane du
Trient

cabane
d'Orny

AIGUILLES ROUGES

refuge
Albert 1er

aig. du Tour ①

plateau
du Trient

refuge du
Lac Blanc

ARGENTIÈRE

glacier du Tour

Tête Blanche ②

glacier de Saleina

cabane
de Saleina

l'Index

La Flégère

LES TINES

③ Chardonnet

SWITZERLAND

④ aig. d'Argentière

le Brévent

LES PRAZ

Montenvers

Grands
Montets

⑤

glacier d'Argentière

refuge
d'Argentière

CHAMONIX

chalet-hôtel
du Plan de
l'Aiguille

aiguille
Verte

Charpoua

Droites ⑥

⑦ Courtes

Dolent

pointe
Isabella ⑧

glacier de Pré de Bar

Mer de Glace

aig. du Moine

glacier de Talèfre

refuge du
Couvercle

Plan de l'Aiguille

Grépon

refuge de
l'Envers
des Aiguilles

glacier de Leschaux

refuge de
Leschaux

⑨ aiguille de
l'Éboulement

tunnel du
Mont-Blanc

aiguille
du Plan

S HOUCHES

glacier des Bossons

⑫

refuge du
Requin

Vallée Blanche

Grandes
Jorasses

aiguille
du Midi

⑬

téléphérique de la Vallée Blanche

glacier du
Géant

Rochefort

⑩

VAL FERRET

refuge des
Cosmiques

Tête
Rousse

refuge des
Grands Mulets

⑭

Tacul

PLANPINCIEUX

Nid
Aigle

refuge du
Goûter

Maudit

pointe
Helbronner

refuge
Boccalate

ITALY

⑰ aiguille de
Bionnassay

⑮ mont
Blanc

Tour
Ronde ⑪

refuge
Torino

LA PALUD

refuge
Durier

⑯

glacier de la Brenva

tunnel du
Mont-Blanc

ENTRÈVES

refuge
Gonella

⑲ aiguilles de
Tre-la-Tête

glacier du Miage

refuge
Monzino

COURMAYEUR

0 Lex Blanche

glacier
de la Lex Blanche

lac de
Combal

Cantine de
la Visaille

VAL VENI

La Thuile
col du
Petit-Saint-
Bernard

Aoste

✳ FOREWORD

Maps
The whole of the Mont Blanc range is covered by two IGN maps: CHAMONIX 3630 OT and SAINT-GERVAIS 3531 ET, TOP 25 series, scale 1:25000. The contour spacing varies along the frontier zones and is indicated on the map used. In Switzerland and Italy the spacing is 20m and in France 10m.

Starting Point
This is the starting point of the climb, from the hut or cable-car (for approach from the valley, see p.94).

The climbing possibilities in the Mont Blanc range are seemingly inexhaustible. Each year, new routes are created, others join the ranks of the classics, and some fall into disuse. The thirty or so routes described here are ideal for becoming attuned to the magnificent Mont Blanc Range. They are on snow or mixed ground and are generally no harder than AD in standard. Distributed among the main glacier basins, they will enable you to visit a representative selection of summits along the full length of the range. Among them are some of the finest classics, which have enjoyed the same popularity down the years.

Yet this selection has an urgent modern emphasis as the mountain environment itself is changing due to global warming, which is affecting routes considerably.

Peaks of medium altitude (up to 3500m) are particularly afflicted. Some of the granite faces have been collapsing spontaneously because the ice that bonds them has disappeared. Some slopes have turned into ice before melting and revealing a base of unstable scree; glaciers have receded and some bergschrunds have widened.

Aiguille Verte with the Drus in the foreground. On the Petit Dru, the rock-falls due to warming at altitude are clearly visible on the photos taken at a ten-year interval: 1995 and 2005.

The south-east side of Mont Blanc du Tacul and its rocky satellites, with the Aiguille de Midi in the background. ➤

Difficulty
It is just as important to be able to judge the context of a climb according to the risks associated with high altitude as to know the technical ability required to accomplish it. These two criteria are taken into account in the two-tier grading system:

Seriousness grade, indicated by a Roman numeral from I to VII, takes into account the seriousness, commitment, accessibility, length, approach and descent difficulty, continuity, number of difficult pitches, gear in place, objective dangers. Any or all of these may define the overall grade:

I short route, accessible, easy descent.
II longer or slightly more technical, descent that may require some care, little objective danger.
III long, sometimes remote, tricky descent, possible objective dangers.
(IV – VII climbs in these categories are not covered in this guide).

The Mer de Glace – the approach to the Couvercle Hut.

The art and practice of alpine sports is thereby changed. Activities are no longer confined to the summer alone and hinge on two linked factors: ground conditions and weather. More than ever, one has to study snow distribution closely, to plan as conditions permit. You choose a route guided by the state of the mountain and not because of some long-held ambition or guide-book recommendation. Even at the height of summer, a big snowfall can sometimes fill up a gully or a bergschrund and the snow can thaw and re-freeze overnight. The only rule is choose the right conditions for your climb. Ensure the snow is firm and stable with the temperature below zero, start at night, and take the safest variants even if they are harder.
Before starting, study maps and guide-books, get information from professionals or places such as the Office de Haute Montagne in Chamonix, take note of the weather forecast and get up-to-date reports from climbers who have just done your proposed route.

Comradeship of the rope on the Dômes de Miage.

The Trient plateau.

The technical grade, from **F** to **D**, then from **1** to **7**, describes the most difficult sections and the hardest pitch.

F (facile): no technical difficulty
PD (peu difficile): requires competence in use of crampons, ice axe and belaying of the leader or second
AD (assez difficile): route with steep sections (45°–50°)
D (difficile): sustained high-angle slope with steep sections (50°–60°), requiring safe technique and a good knowledge of belaying
1: long sections at 60°
2: sections at 70° but good belays
3: sections at 70–80° generally on good ice. Steep sections alternate with good resting places and belays
4: sections at 75–85° sometimes with a short vertical section. Ice generally good with good belays.

Time Given for a party with members of equal technical standard, competent at the grade, in good conditions. Keeping to time is an important safety factor. If your progress is slow be prepared to retreat, before the point of no return.

Favourable conditions
In this high massif at the exposed south-western end of the Alps, weather conditions are subject to constant change. One must learn to observe and make judgements, read and interpret. You can climb all year, as long as conditions are good: enough snow, stable slopes, correct temperature …

Equipment Specific equipment for a party of two is indicated. Basic personal equipment is also required.

Other items
rope we state the length necessary for abseils or the minimum distance between belays.
ice pegs should never be absent from the gear-rack when on a glacier (in case of a fall into a crevasse); take two per person.
nuts usually a full set from 1 to 10.
friends 2, 2.5 and 3 are best.

Note for UK / US Edition
Most climbers will have both technical axe and ice hammer (with curved or banana picks), though the easier climbs could be done with less. Ice pegs (or screws) are per party assuming, for ice pitches, a minimum of two per belay and two for running belays. A few rock pitons might be taken, for though nuts will usually suffice a solid piton to secure an escape abseil is often valuable. It is assumed that all climbers will have helmets, headtorches, goggles and glacier cream. Abseil lengths for descents or bergschrunds may suggest taking both a normal rope and a lighter abseil rope.

TRAVERSE
TABLE COULOIR

Starting point   Albert Premier (1er) Hut (2702m): 1hr 30mins from the Col de Balme. Approach p94.

Keen alpinists, hurry up and climb the Aiguille du Tour! If climate change continues, the descent from the Normal Route may terminate in a gaping bergschrund, unless someone puts a step-ladder up it! Mountaineering these days is done on an opportunistic basis. You need to watch the weather closely and know when to start when the time is ripe.

The traverse suggested starts from the Albert Premier Hut and has the advantage of a rarely used approach by the Col du Midi des Grands that is less frequented than the traditional southerly approach. The surroundings are wilder and the route crosses two successive cols to the north of the peak with fine views of a high-altitude landscape looking across to the nearer parts of the Swiss alps. Once you are on the Trient plateau, don't miss the opportunity to visit the Fenêtre du Pissoir, with its secret charm. One then crosses the Trient Glacier below the North Peak rather than crossing its summit as the rock of its North Ridge is now too loose, particularly on descent. Several ways of climbing the South Peak are now possible, depending on the state of the

bergschrund and the queue!

The summit platform, suspended between Switzerland and France, offers an astonishing panorama, the stuff of innumerable future adventures. Regarding suitable conditions – although these lower peaks lose their snow rapidly when it gets hot, they also allow very rapid ascents if there is snow. A snowfall in mid summer sometimes consolidates in twenty-four hours. So the Table Couloir, facing south-west, can be climbed if you make sure that there is stable consolidated snow, that has been re-frozen overnight. The uniform-angled slope barely steepens before giving out on to the gap above the pinnacle that bears the big rock table, which gives the route its name and is easily seen from the hut. Above this follow the rocky ridge, usually mixed at the start of the season and quite exposed, with some blocks that require care. The route along the ridge permits some variation so as to take advantage of the natural protection offered by passing the rope behind a rock spikes. It is best to follow the Normal Route.

TRAVERSE

Difficulty   II/F, glacier travel (slopes of 40°) and mixed (easy rock), bergschrund sometimes tricky.
Time   From hut to summit, 3–4hrs; descent 1½hrs.
Height gain   840m.
Conditions   Better at the start of the season. Otherwise, make sure there is enough snow.
Equipment   Crampons, axe, rope (possible 50m abseil at the bergschrund), 2 ice pegs, 3 slings, 3 karabiners.
First ascent   C.G. Heathcote and M. Andermatten, 18 August 1864.

Approach climb by the Col du Midi des Grands
From the Albert Premier Hut, follow the route eastward (boulders, snow, cairns), to the Tour Glacier (20mins). Start up east and at the top of the steep slope which goes round the Signal Reilly (2883m), go straight up to the north-east to climb the left bank of the valley which opens

up on the left of the West Ridge of the Aiguille du Tour. It leads out on to the Col du Midi des Grands (3235m). Climb the slope (sometimes rocky) diagonally from right to left. Traverse east to reach the gap between the Aiguille du Pissoir and the Pissoir, which enables you to set foot on the Trient plateau and

On the Traverse,
above the Pissoir.

The two Aiguilles du Tour
and the Trient glacier seen
from the Tête Blanche.

aiguilles du Tour
South summit   North summit   le Pissoir
3542   3544   3319

aiguille
Purtscheller
3478

aiguille
du Pissoir
3441

col Supérieur
du Tour
3289

plateau du
Trient

AIGUILLE
DU TOUR
East Face

the eastern flank of the mountain. From there, traverse south up the snow slopes below the Aiguille du Pissoir, passing below the North Summit of the Aiguille du Tour to join the Normal Route to the South Peak. 2½hrs from the hut.

An interesting variation in the approach is possible in good, stable snow conditions (more likely at the start of the season). From the Col du Midi one can climb the west aspect of the Fenêtre du Pissoir (100m, a little harder than the more northerly route, 35°–45°). This short cut brings one to 3410m on the Trient glacier, between the Aiguille du Pissoir and the North Peak of the Aiguille du Tour.

Ascent
Cross the bergschrund and gain the rock where crampons/axes can be left if it is clear of snow.

Three ways are now possible:

(a) Follow the natural ledge which crosses diagonally the whole of the east side of the Aiguille. It leads to a small gap on the North-East Ridge where, after a little mantelshelf, the summit is soon reached (20mins).

(b) Climb the short gully between the two peaks (keeping to the east side), then take the crest of the ridge on the left to the summit (25mins).

(c) Follow the crest of the ridge as directly as possible, on its right (moves of 3b, sometimes steep and exposed, but with good holds (30mins).

Descent by the Col Supérieur du Tour
Return to the bergschrund and from there, heading south-east then south, cross the slopes of the Aiguille Purtscheller (3289m) – steep for 50m – to the Col Supérieur du Tour. Cross it heading west (a few metres on rock). Descend the wide and fairly steep slope leading to a basin (3120m) at the foot of a fine reddish, granite buttress. After a slight rise to the north-west, the approach route is rejoined. Go down the big slope north of the Signal Reilly and follow the path to the hut. 1½hrs from the summit.

Note The traverse can be done in both directions and each route can obviously be done as a return trip. The Aiguille du Tour also may be done from the Trient hut, crossing the plateau of that name.

TABLE COULOIR
Difficulty II/AD–. Snow gully at 40°–45°, then an airy rock or mixed ridge, depending on the time of year.
Time Approach 2hrs. Ascent 2hrs. Descent 1½hrs.
Height gain Hut to summit 840m (300m for the couloir).
Conditions Being south-facing ensure the snow is well frozen.
Equipment Crampons, ice axe, one hammer-axe per party, rope (possible 50m abseil for the bergschrund on the east side), 2 ice pegs, a selection of slings, 3 quickdraws.
First ascent Date and party unknown.

Approach After climbing for 20 minutes on a very rough path (boulders, snow patches, cairns) reach the Tour Glacier. Go up (eastward) passing to the north of the Signal Reilly (2883m). After traversing south-eastwards, go diagonally south, then go up more steeply. As you approach 3100m, turn sharply north-east so as to enter (outflanking big crevasses) the glacier basin under the south side of the Aiguille du Tour. Go up northwards to reach the foot of the couloir, which is now obvious. 1½hrs–2hrs. Colour sketch p15.

Ascent Start toward the right-hand side and soon cross to the left of the channel which

usually runs from top to bottom. Go up the right bank (to the left looking up). The slope steepens a little before leading to a gap at 3529m on the West Ridge (loose blocks, take care not to cause stonefall). 1hr. Go up right to reach the summit by the ridge, rock or mixed. 3–4hrs from the hut.

Descent on the east side, by the Normal Route
Climb down the North-East Ridge for 60m to reach a little gap, then follow a good ledge to the right (this slants across the East Face) and after a short slope, cross the bergschrund. From there, reach the Col Supérieur du Tour and the hut, by the route previously described.

Àt the exit from the Table Couloir with the Table Gendarme in the background.

brèche 3529

gendarme
à la Table

aiguille du Tour
sommet Sud
3542

aiguille
Purtscheller
3478

Pain de Sucre
du Tour

col Supérieur
du Tour
3289

glacier du Tour

• 3120

AIGUILLE DU TOUR
South Face

NORTH FACE
AND TRAVERSE

Start   Albert Premier Hut (2702m) or Trient hut in Switzerland. Approach p94.
Difficulty   Tête Blanche I/AD, short steep slope 120m at 50°–55°. Petite Fourche F, mixed.
Time   Approach 3hrs, North Face 1hr, Petite Fourche 1½hrs out and back, descent 1hr.
Height gain   720m + 100 m for the Fourche.
Conditions   Best June–August, before the ice disappears and the bergschrund becomes too big.
Equipment   Crampons, ice and hammer axes, 50m rope, 6 ice pegs (two per stance), 3 slings, 3 quick draws, 5 karabiners.
First ascent   Tête Blanche North Face (known).
            Petite Fourche: E. Dufour, H.-R. Whitehouse with H. Copt, 16 August 1876.

Organising crampon and ice-axe practice, and placing ice pegs is becoming ever more difficult as the glaciers retreat. The head of the Trient Glacier is now a valuable site. Here you can quickly pick up the basic skills and put them to use on a real climb.

This mountain excursion combines two short routes, with a modest level of commitment, and provides an opportunity to try out or perfect skills that are sure to be required. For these purposes the Tête Blanche is ideal, for apart from its easy Normal Route, it also has a short, sharp North Face, ideal for higher-standard training.

Judging suitable snow and ice conditions for a climb of this type is the first test. These are best when the slopes are hard snow with patches of snow/ice stabilised by a melt-freeze cycle.

The bergschrund, where the glacier splits from the face, easily crossed when full of snow, becomes far harder when open and the slope above icy. Using technical axes and front points, belaying on ice pegs, perfecting ice-face rope handling will all reinforce confidence.

If you started in the early hours, the first rays of the sun will greet you at the top. Afterwards you can carry on to the pleasant summit of the Petite Fourche, which will lift you up a step above the Fenêtre de Saleinaz, opening on to the vast horizons of the Valais where the Matterhorn, Dent Blanche and Weisshorn dominate the distant horizon.

A bergschrund on Tête Blanche, Aiguilles Dorés beyond.

Petite Fourche (left) and Tête Blanche with its North Face and bergschrund on the right.

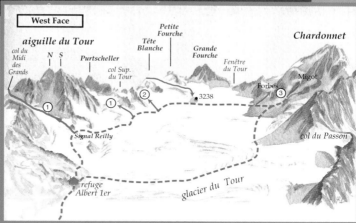

The Albert Premier Hut and Aiguille du Chardonnet (left).

Given the north-east orientation of the face, it is essential to be at the bottom of the face at dawn.

Approach From the Albert Premier Hut, go east by the intermittent path (20mins) to gain the Tour Glacier. Go up it eastward, passing north of Signal Reilly (2883m). After a traverse followed by an easement, climb south-east up a steep slope, then, following the 3100m contour, reach a basin at the foot of the gully coming down from the Col Supérieur du Tour (Normal Route for the Aig. du Tour). Continue, passing the foot of the spur (3131m) coming down from rock needles at the Col du Tour, and head east up the slopes leading to the Col. Keep left at the end, following a fairly prominent snow ridge, to come out to the north of the Col at some rocks (a deep basin prevents direct access on the right). 3282m, 2½hrs.
Descend to the Trient plateau and traverse about 200m to the south-east, under the North Face of the Tête Blanche. This trapezium-shaped face is bounded on the right by the North (Frontier) Ridge, which is mixed, and on the left by the snowy North-East Ridge coming down from the summit.

Ascent It is usually easier to cross the bergschrund on the left, almost directly below the summit. Climb the steep slope (50°–55°) to the top of the Tête Blanche (1hr, 3–4hrs from the hut). The North and North-East ridges mentioned above also provide fine routes, of interest especially if conditions (bergschrund, unstable snow) prevent an ascent of the North Face.

Descent by the Normal Route (west side)
From the summit, make a wide arc (south, then west) to avoid the long bergschrund which bars a westerly descent. After passing the rognon at 3238m, go due north to reach the flat area at the foot of the Col Supérieur du Tour and by the ascent route to the hut. (1½hrs. Colour sketch, p15.)

Petite Fourche
From the summit of Tête Blanche go south to the Col Blanc (3405m). Go up the snow slope to the crest, to the right of the summit of the Petite Fourche and follow easy rocks, sometimes with snow, to the highest point (3520m). 1hr. Go back by the same route to Col Blanc and from there by the west side, join the Tête Blanche descent.

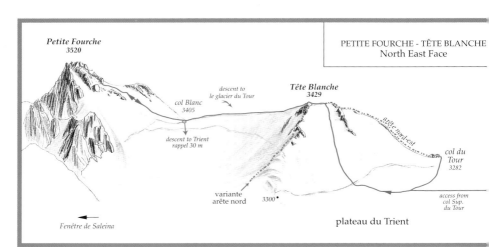

Petite Fourche
3520

PETITE FOURCHE - TÊTE BLANCHE
North East Face

descent to
le glacier du Tour

Tête Blanche
3429

col Blanc
3405

arête nord-est

descent to Trient
rappel 30 m

col du
Tour
3282

variante
arête nord   3300

access from
col Sup.
du Tour

Fenêtre de Saleina

plateau du Trient

The 100m North Face of Tête Blanche.

AIGUILLE DU

FORBES ARÊTE
MIGOT SPUR

Start Albert Premier Hut (2702m), 1½hrs from the Col de Balme. Approach p94.

Some mountains flaunt their character appearing indestructible, as difficult to descend as to climb. The summits merely represent a stage in the climb as getting back to the valley is often as hard as the route itself. So it is, for example, with the Jorasses, the Verte ... and the Chardonnet.

The Forbes Arête is thus considered one of the classic finest climbs in the Alps with open climbing and splendid views.

The route has an innate logic – everything well laid out – the night-time glacier approach, the abrupt start up the lower slopes, the exposed Ice Bosse requiring sound technique on slopes up to 50° that can present bare ice later in the season.

Once reached, the Arête is impressively narrow and requires climbers to be of equal ability. It is sometimes corniced and has icy mixed ground where crampons are retained, even on the pure rock sections, as they often alternate with ice. For speed

with safety, moving together will be necessary. Nearing the summit, on the right, the Migot Spur strikes the eye: perfectly direct, supporting the summit itself, a powerful pillar with a harmony and allure that is bound to bring you back on another day.

The Migot Spur has *grandes courses* pretentions. You will meet stimulating terrain: an initial slope done by headlamp, rocks mixed with ice, lit by the morning sun, the final slope without a single blemish. It's a big effort and with body sated, one would easily be content with a humdrum summit. No such anti-climax here – while the body rests and is restored, you can admire the profusion of north faces above Argentière, the fortress Jorasses and the cyclopean bulk of Mont Blanc. And how sharp our summit is! So much beauty giving so much happiness.

The descent remains and that too, with less snow cover, has become harder in recent years.

Aiguille du Chardonnet seen from Petite Fourche – the Migot Spur is on the right and the Forbes Arête the left skyline.

Dawn: the ideal time to arrive at the Forbes Arête of the Chardonnet.

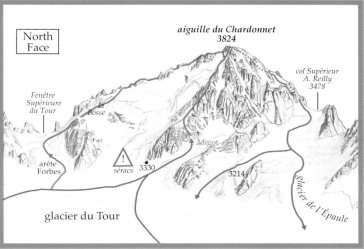

North Face

aiguille du Chardonnet
3824

col Supérieur
A. Reilly
3478

Fenêtre
Supérieure
du Tour

la
Bosse

Migot

arête
Forbes

séracs 3330

3214

glacier de l'Épaule

glacier du Tour

Forbes Arête
Difficulty   III /AD   Mainly glacier terrain with some mixed. The descent is increasingly tricky
as the day goes on and can be difficult if icy.
Time   5–6hrs hut to summit. 2–3hrs for the descent.
Height gain   Approach 550m, bergschrund to summit 600m.
Conditions   June – September : good snow cover and cold conditions required.
Equipment   Crampons, ice axe and short axe, rope for a 30m abseil, descender, 4 ice pegs,
selection of slings, 5 quickdraws, 4 karabiners.
First ascent   P. Sisley with M. and J. Crettex, 10 August 1899

On the Forbes Arête with Aiguille d'Argentière behind.

Approach   From the hut, take the classic
Aiguille du Tour route keeping high on the
east side of the glacier (east, then south-
east). When below the small valley leading
to the Col Supérieur du Tour, head south,
gradually rising to pass just to the west of
the rognon at 3238m; after passing it, go
across more or less on the 3220m contour,
aiming for the bottom of the wide couloir
with séracs, bounded on the left by the
Aiguille Forbes and on the right by a rocky
rognon. 2hrs. Colour sketch p15.

Ascent   Climb the couloir (crevasses) to
reach a glacier bay (big gendarme on the SE).
Cross the bergschrund on the right (west) to
gain access to the Ice Bosse. Climb this (50°
for about 100m). Continue up the easy slope
above to gain the East Ridge at a gap at
around 3700m. Climb the first rock outcrop
(keeping crampons on), working more to the
south side, then follow as closely as possible
the summit ridge, which is a mixture of snow,
ice and rock steps. There is nothing too

difficult, if you follow the obvious route
markings – crampon scratches, ledges and
gullies on right or left – thereby avoiding
the few difficulties presented by some rock
pinnacles. Go down a short snow slope to the
south of the last pinnacle to tackle the mixed
ground of the summit rise. Just after the top
of the Migot Spur (the fine slope to the
north), the summit is reached. 3–4hrs, 6hrs
from the hut.

Descent on the west side   Go down the
easy South-West Ridge to the exit of the
obvious West Gully. Descend this for about
150m to a rocky rognon and another snow
slope. Then descend keeping left (west,
avoiding a steep gully which plunges down
on the Argentière side), to reach the start
of a secondary gully, quite steep and narrow
(rock, sometimes verglassed with stonefall
risk, possible anchor on the right for a
30m abseil). This leads after 100m
to the snowy Col Supérieur Adams Reilly.
Go down the steep slopes of the
north flank of the Glacier de l'Épaule
(bergschrunds sometimes high and wide,
observable from the hut; an abseil stake
may be useful).

To reach the Tour Glacier, it is generally
better to pass to the right of the rocky rognon
which divides the lower part of the Glacier de
l'Épaule, but if the crevasses are open, it is
possible to go left of the rognon. In any case,
one has to tack back and forth in an area of
big longitudinal crevasses. If the Tour Glacier
in well snowed-up, it is possible to head
fairly directly west, then due north towards
the hut; otherwise, go up north-east to join
the approach route. 2–4hrs from the summit
depending on conditions.

Migot Spur

Difficulty IV/AD+ Mainly snow with around 100m of mixed ground, where one should keep to the right (good protection). The final slope of 50° is quite exposed. The descent (west facing) can be problematic if icy.

Time Approach 2½hrs. Ascent 3–4hrs. Descent 2–3hrs.

Height gain About 680m on approach and 450m on the climb.

Conditions Good freezing conditions and snow cover for the first part and the mixed ground and the final slope without too much grey ice. Usually found in spring or early summer. The descent rapidly becomes icy – characteristics related to the slope orientation: north-east for the route, west for the descent.

Equipment Crampons, 2 technical axes, rope (possible 30m abseil), 6 ice pegs, deadman, 5 quick draws, 4 karabiners, slings, nuts, Friends 0.5 to 3.

First ascent C. Devouassoux, 28 July 1929.

Approach There are two ways to approach the bergschrund:

If the glacier has good snow cover, leave the hut on the south-east side and cross the glacier southwards, turning some big crevasses on the right (around 2850m). After passing them, head south-east going towards the foot of the spur (1hr 20mins). Alternatively, if the crevasses are open, make a detour by the upper part of the glacier (north) from the hut, following the approach to the Forbes Arête route to the rognon at 3238m. From there, head south-east along the 3200m contour to the foot of the spur (2hrs – don't loiter as there is sérac fall danger). Here the two approach routes join. Climb the snow bay just to the right of the buttress and follow its west flank until below the lower snow ridge which marks the start (c.3380m) of the Migot Spur at a bergschrund. (30mins, 2–2½hrs from the hut).

Ascent Go up the snowy slopes, sometimes mixed, to the ridge. Follow it south and continue up the slope into which it merges, before arriving at the foot of the mixed ground. Climb this (3 or 4 pitches) keeping as far as possible to the right to avoid any sérac falls. Continue up the rocks to the upper snow ridge (bergschrund) and finish up the fine slope (100m 50°–55°), exiting to left or right of the summit rocks. 3–4hrs.

Descent see p20.

The Migot Spur.
Two parties can be seen on the final slope. ➤

NORMAL ROUTE
WHYMPER ROUTE

Y COULOIR
FLÈCHE ROUSSE RIDGE

Start Argentière Hut (2771m), 1½hrs from the Col des Grands Montets. Approach p94.

The Aiguille d'Argentière is a large and complex peak. It resembles one of those huge Oriental fortresses, its bulky features somehow attaining a subtle harmony despite the sheer architectural scale. This semblance of order is assisted by the interspersing of snow slopes amidst the great pillars of yellow granite that radiate from the rounded snow summit like the rays of a starfish. The routes are varied and aesthetic, but always serious because of their length.

We begin with the traverse, which after a long approach, climbs the last two hundred metres of the steep North Face, a fine first ascent bearing the name of Whymper. The descent by the Milieu Glacier turns it into an expedition of some commitment requiring good movement skills, front pointing, and the competent use of axes and ice pegs. Those seeking something rather less taxing can go to the summit and back by the Milieu Glacier, seeking a reasonable bergschrund crossing and finding an ice-free line up the final slope.

An unusual feature for a southerly aspect, the snow-filled Y Couloir, slopes at a uniform angle for 500m and there is no need to force the pace to gain the welcoming rounded summit. The slope aspect means that the ascent is best completed before sunrise. Stay alert for stonefall in the upper part by looking up frequently and pushing the ice axe in firmly at each step.

The Flèche Rousse Ridge offers complex terrain where following the exact crest of the ridge is impossible. It is necessary to assess what difficulties lie ahead, climb in pitches and zigzag from one side to the other – tackling rock walls, crossing icy chutes, bridging gaps, making short abseils and dealing with slippery rock and rock needles. A good stiff mountaineering work-over.

Whatever your choice from these four diverse routes you will be travelling in wild terrain where a vast and magnificent landscape gradually unfolds, nearer and nearer to the sky, far from the beaten track.

The Argentière Glacier – Aiguille du Chardonnet (left), Aiguille d'Argentière (centre), Tour Noir (right).

D'ARGENTIÈRE (3902m) ARGENTIÈRE

The summit view south-east from Aiguille d'Argentière.
Tour Noire and Mont Dolent are in the background.

The Argentière Hut.

from the west

Chardonnet

aiguille
d'Argentière

Tour
Noir

glacier
du
Chardonnet

c
d
a
refuge
d'Argentière

b

glacier d'Argentière

moraine des Rognons

Lognan

Grands
Montets

(a) Normal Route by the Milieu Glacier

Difficulty III/PD+ A steep glacier, quite open with an often icy summit slope (400m at 40°–45°).
Time Ascent 4hrs. Descent 2–3hrs. **Height gain** 1130m
Conditions Stable snow conditions needed. Before departing check the bergschrund size and the condition of the summit slopes, which quickly become icy.
Equipment Crampons, ice axes, 50m rope (useful if pitching higher up), 6 ice pegs (if final slope is icy), selection of slings, 5 quick draws, 4 karabiners.
First ascent L. Dècle, Y.A. Hutchinson with A. Imseng and L. Lanier. 14 August 1880 (in descent).

Approach and ascent From the hut, take a good path going north-west and leading quickly to the moraine on the left bank of the Milieu Glacier. Follow the moraine crest. Cross under the glacier tongue (north-west, stone and ice-fall danger), to gain the right bank of the glacier, which is followed to a small plateau. Go back to the left bank and climb to the bergschrund guarding the summit slope. After a steep ascent of 40m, turn right, after which the summit is soon reached. 3–4hrs from the hut. Descend by the same route.

The top part of the North Face, at the exit of the Whymper route.

(b) Whymper Route and Traverse

Difficulty   III/AD— A long snow traverse, with a 35°−40°gully and a north facet at 40°−50°.
Time   Ascent 5hrs. Descent 2−3hrs.
Height gain   1130m.
Conditions   Re-frozen snow, overnight frost.
Equipment   Crampons, ice axes, 50m rope (useful if pitching higher up), 6 ice pegs, selection of slings, 5 quick draws, 4 karabiners, nuts.
First ascent   A. Adams Reilly and E. Whymper with M. Croz, M. Payot and H. Charlet, 15 July 1864.

Approach and ascent   From the hut, by the route of ascent, descend the Argentière Glacier to around 2570m. Then traverse NE to the moraine on the left (south) bank of the Chardonnet Glacier (cairns, 45mins). Climb the moraine and move onto the glacier at about 2750m (40mins). Follow the left bank then, after going left round the slightly steeper middle part of the glacier, quit the Col du Chardonnet line and go diagonally right. Then climb the south branch of the glacier, which steadily steepens to the foot of a very obvious couloir coming down from the NW Ridge of the Aiguille d'Argentière (2hrs from the top of the moraine). Cross two bergschrunds (around 3600m) and after about 150m in the couloir (35°) emerge onto a snow (view across the Saleina face). Climb up the right-hand side of North Face of the Argentière by the fine ridge crest (150m at 45°−50°), to reach the main summit (1hr from the shoulder, 5hrs from the hut).

Descent by the Milieu Glacier   Descend the south-west slope (400m at 40°−45°, delicate if icy). At first follow the left bank of the Milieu Glacier, then as it opens out (small plateau), cross to the right bank and descend the tongue. Fifty metres below the tongue, head south-east keeping alert for ice and rock falls. After crossing the moraine on the left bank, reach the hut by a good path (start marked by a cairn). 2−3hrs from the summit. Alternatively (to avoid going back to the hut) descend the moraine directly to the Argentière Glacier from where you can descend to Lognan (2hrs from the foot of the Milieu Glacier).

(c) Y Couloir

Difficulty   III/AD A uniform 45° snow couloir with a sometimes awkward bergschrund (3b variation by rocky rognon). Descend by the Milieu Glacier (400m at 45°) steeper at the top
Time   Approach 1½hrs. Ascent 2−3hrs depending on the bergschrund. Descent 2hrs.
Height gain   1130m, including 450m in the Couloir.
Conditions   Because of its aspect, the snow is usually stable after a good overnight frost.
Equipment   Crampons, axes, rope, 4 ice pegs, slings, 3 quick-draws, 4 karabiners, nuts.
First ascent   H. Cameré, solo, in July 1922.

Approach   Aim to start the Couloir before dawn. From above the hut, head east (path by the water-pipe) and go up the moraine to the north-east. Gain the Améthystes Glacier (45mins) and follow its right bank, below the Minaret, to reach the bottom of the Y Couloir (3300m). 1½hrs from the hut.
Ascent   Cross the bergschrund on the right if it is not too open (stonefall risk). Otherwise, start lower on the right, by the rocky rognon which borders the Couloir — climbing a steep, left-slanting chimney and thence up left (fractured rock) to enter the Couloir (a few moves of 3b). Cross the debris runnel to climb the right bank. After passing the left branch, carry on up the middle of the right branch to reach a prominent col on the ridge, to the left of the rock buttress of the Flèche Rousse. Head north-west, by the slope bordering the ridge and soon, arrive at the summit. (2−3hrs from the bergschrund). Diagram p27.
Descent by the Milieu Glacier see above.

The Tour Noir Glacier (in shadow) and (beyond) the Améthystes Glacier slant up right from the upper Argentière Glacier.

(d) Flèche Rousse Ridge

Difficulty III/AD A lengthy mixed route with snow at 50°–55° and 4b rock sections. Requires adroit rope-handling and shrewd route-finding. Descent by the Milieu Glacier.
Time Approach: 2½hrs. Ascent: 3–5hrs. Descent: 2hrs.
Height gain 430m bergschrund to summit.
Conditions A climb where some extra snow cover (stable – after a clear, cold night) is a big help.
Equipment Crampons, axes, rope for 25m abseil, descender, 4 ice pegs, 3 quickdraws,
3 karabiners, slings including 3 long ones, nuts, Friends 2 and 3.
First ascent G.H. Morse, J.H. Wicks and C. Wilson, 3 August 1893.

Approach From above the hut, follow the water-pipes north-east. On an intermittent path, go up to the right and gain the crest, then the top, of the Améthystes Glacier moraine, reaching the glacier itself around 3100m (45mins). Go up its right bank, cross under the Y Couloir and continue to the bottom of the first of three gendarmes standing above the Col du Tour Noir on the left (north) which mark the start of the Flèche Rousse Ridge. 2½hrs.
Ascent Start up the slope, which goes north under the wall formed by the gendarmes, to a short gully that leads to a col on the crest of the ridge. After turning a pinnacle on the Saleina side, go up some mixed ground to reach a wide flat area.
Avoid the temptation to turn the steep section above by its right side, and instead, on the Améthystes side, go down a rather forbidding couloir for twenty metres and on the south-east side, go straight ahead up a steep diagonal chimney (30m, delicate, 4b). Ledges lead back right giving access to a wide couloir which runs along below the crest, then angles upward higher up. Take the easiest way up, to regain the crest of the ridge, then by a steep rib on mixed terrain, reach the foot of the Flèche Rousse (Red Arrow). Cross to the Améthystes side, climb some steep but easy rocks, go left around a rock overhang, pass under the big flake that gives the climb its name, and reach the summit over some blocks (3827m, 3hrs). Abseil 20m on the north side down a smooth slab. By terraces, reach a snow saddle (3827m), descend to a wide col (exit of the Y Couloir) and gain the summit of the Aiguille d'Argentière by the easy ridge. 30mins, 4hrs from start.
Descent by the Milieu Glacier see p24.

South Face

*aiguille
d'Argentière*
3902

Flèche Rousse
3832

3618

*col du
Tour Noir*
3535

**arête de
Flèche Rousse**

couloir
Y

glacier des Améthystes

The start of the
Flèche Rousse Ridge.

NORMAL ROUTE
NORTH-WEST FACE
CHEVALIER COULOIR

Start From the Col des Grands Montets (3233m), under the top station of the cable car, that leaves from Argentière. See p96.

The little shoulder that starts the Grands Montets Arête of the Aiguille Verte still gives badly prepared climbers something to think about. Access by cable car, lack of an approach march and the PD standard of the Normal Route, doesn't mean that a certain basic preparation in icecraft can be overlooked. The ridge has all the problems of the high mountains; altitude, exposure, a tricky bergschrund, mixed terrain, patches of ice, rocky sections sometimes climbed with crampons on and moving together. The close-up views of the Verte's formidable Nant Blanc Face and the north side of the Dru are striking. Modest it may be, but the Petite Verte remains an excellent training summit.
Early in the season, the North-West Face provides a good exercise in technique, though it is best avoided if there are more that one or two parties on the Normal Route.

The east-facing Chevalier Couloir, requires a nightime start so that climbers will reach the summit at the same time as the sun first touches it. Otherwise, it may turn into a perilous venture. The interest of the route lies in its diagonal line, across narrow Andean-like ice runnels, constantly moving from one to the other around the narrow ridges that separate them. Two axes per climberare very useful, and the party members should be of equal ability, able to deal with these awkward steps after the little thrill of the nocturnal bergschrund crossing. Climbers will suddenly be made aware of the route's orientation when, almost at the top of the couloir, the sun tints the rocks, warms biceps and calves, but rapidly destabilises the ice. Having quit the couloir, the colder North-West Face leads to the summit.

Normal Route and North-West Face
Difficulty I /PD mixed route, steep slope on to the ridge and some moves of 3b on rock, moving together. — *North-West Face* I/AD, 50°.
Time From the Col to summit, 1½hrs–2hrs. Descent 1hr. **Height gain** 279m.
Conditions May – October: Get information about snow conditions as the slopes can be icy.
Equipment Crampons, ice axes, rope, 2 ice pegs, selection of slings, 3 quick draws, 5 karabiners.
First ascent J.-E. and R. Charlet with P. Charlet, September 1886.
North-West Face A. Contamine, 1959.

Normal Route From the Col des Grands Montets (reached by descending the stairs under the top station of the cable car), go up the wide moderately angled slopes, due south to begin with, then head SW towards the distinctive shoulder on the ridge. Cross the bergschrund on the right (wide and tricky at the end of the season) and come out on the shoulder. After a long section traversing below the crest of the ridge

(many spike belays), join the ridge at an obvious little gap (axes and crampons can be left here). After a little wall (3b) and an easement, you will reach a break in the ridge (the Salle à Manger), about 60m from the summit that consists of rock flakes. To get there descend a steep little wall (3c – easy on re-ascent) and in a few minutes you arrive at the true summit. 2hrs.
Descent 1hr.

◄ The short summit wall.

The top of the Chevalier Couloir.

Petite aiguille Verte
3512 *passage du mur*

PETITE AIGUILLE VERTE
from the north-west

shoulder

facette nord

glacier des
Grands-Montets

**Petite Verte (middle distance) gives a fine view of
Aiguille Verte (high above) and its Nant Blanc Face (right).**

North-West Face
At the easement which follows the initial slope above the télépherique station, ascend head south-east to the bergschrund on the left side of the face. This way avoids any stonefall triggered by parties on the Normal Route.

Once over the bergschrund, bear left under the rocks (two 50m-pitches), before breaking right towards a gap in the ridge (three pitches). The NNE Ridge can also be climbed from the bergschrund.
Descent 1 hour by the Normal Route.

Chevalier Couloir

Start Col des Grands Montets or Argentière Hut. Approach p94.
Difficulty II/AD+ to D depending on conditions, particularly at the bergschrund. 50°–55°.
Time Approach 3hrs from the hut, or 20mins from a bivouac on Col des Grands Montets.
       Ascent 3hrs. Descent 1hr.
Conditions Spring and very early in the season after that, the snow disappears! Leave very early.
Equipment Crampons, ice and hammer axes, 6 ice screws (per party), a dead man, slings,
3 quickdraws, 4 karabiners, nuts.
First ascent H. Cameré and P. Chevalier, 23 August 1930.

It is absolutely essential to start the couloir before sunrise and in good freezing conditions.

From the Argentière Hut
Cross the glacier so as to descend on its left bank and at c2600m, bear diagonally to the left to go up in a westerly direction following the Rognons moraine. At about 3050m (shortly after a little rock monolith), ascend diagonally left to gain the foot of the couloir. 3hrs.

Col des Grands Montets Start Descend eastwards (bergschrund) and go right along the bottom of the north-east side of the Petite Aiguille Verte, to the foot of the couloir. 20mins.

Ascent Cross the bergschrund and go diagonally right up the slope, from runnel to runnel to the col where the couloir comes out, keeping somewhat to the right (do not be drawn towards a steep section directly in line with main gully). The exit from the couloir is sometimes mixed and can be solely on rock. You are now on the edge of the North-West Face of the Petite Verte, which is climbed in a few pitches at 50° to the summit. 3hrs from the bergschrund.

Descent Take the West Ridge (Normal Route) to the Col des Grands Montets. 1hr. Diagram p29.

Chevalier Couloir in the dawn sun (centre/left) with the Col des Grands Montets (right).

Traversing from runnel to runnel near the top of the Chevalier Couloir.

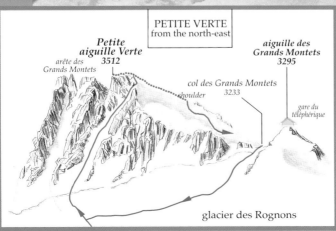

PETITE VERTE
from the north-east

*Petite
aiguille Verte*
3512

*aiguille des
Grands Montets*
3295

*arête des
Grands Montets*

col des Grands Montets
3233

shoulder

*gare du
téléphérique*

glacier des Rognons

NORMAL ROUTE: SOUTH FACE BY THE EASTERN SPUR

Normal Route: South Face by the Eastern Spur
Start Couvercle Hut (2687m). Approach p94.
Difficulty II /AD+, mainly on snow, with an approach couloir and slopes of 45°.
Time Approach 2hrs. Ascent 3–4hrs. Descent 3hrs.
Height gain 600m from bergschrund to summit.
Conditions Snow on south-east slopes melts and re-freezes quickly, favouring early-season ascents. The rock sections should be clear of snow, but bergschrunds soon widen. The descent slopes quickly become avalanche-prone, so an early start and keeping to the timetable is important.
Equipment Crampons, 2 axes, 50m abseil ropes essential, descender, 4 ice pegs, deadman, 5 quickdraws, 5 karabiners, slings and abseil slings, nuts, Friends 2 to 3, a few pitons.
First ascent H. Cordier, T. Middlemore, J. Oakley Maund, J. Jaun, A. Maurer, 7 August 1876.

In the same way as the Grandes Jorasses, the Pilier d'Angle or the Drus, Les Droites inspires respect. Whether one is an experienced climber, or just beginning an alpine odyssey it is impossible not to be aware of these great peaks whose fine climbs have become established cornerstones in the history of alpinism.

The final ridge of Les Droites high above the Talèfre Glacier.

Standing above the Argentière Glacier, the North Face of Les Droites presents an inspiring blend of snow and ice drapery sweeping up to powerful granite buttresses each divided by menacing ice couloirs. A clutch of fine *grandes courses* work up this difficult 1000m face: climbs coveted by many ambitious alpinists.

The southern (Talèfre) side of Les Droites is more amenable. The peak has two summits: that on the west at 3900m, the easterly one measures exactly 4000m. By good luck, a huge pillar falls directly from the little snowy tooth at the highest point all the way down to the glacier 1000m below. This is the eastern spur that we recommend. It is a demanding climb with route-finding and technical aspects that require efficient mountaineering skill and speed to complete within a safe time-scale.

It is gained by its south-west couloir that accesses the crest of the spur, and then goes up a rock wall where 4c climbing combines with the icy, adventurous terrain. The finish is up a snow ridge stretching out towards an ideally airy summit. As you gain height, the whole of the southern aspect reveals its secret treasures: the Courtes, Triolet and Talèfre encircle the Talèfre Glacier and beyond the Grandes Jorasses dominates. To the west is the stately Aiguille Verte, while to the south-west Mont Blanc looks magnificent.

The summit view from
Les Droites to the
Grandes Jorasses
and Mont Blanc.

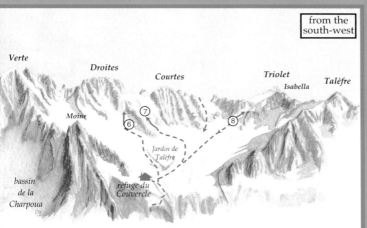

Verte

Droites

Courtes

Triolet

Talèfre

Isabella

Moine

⑦

⑥

⑧

Jardin de
Talèfre

bassin
de la
Charpoua

refuge du
Couvercle

Approach From the hut, descend eastwards by the path to the Talèfre Glacier (2640m, 10mins). Follow the path northwards up the middle of the glacier to 3000m, turning the Jardin de Talèfre on the left. Then bear north-east towards a south-west facing couloir which cuts the left flank of the eastern spur. By a right-to-left turning movement avoid a crevasse, and reach the bergschrund (3400m). 2hrs, additional sketch p33.

Note: If crevasses left of the Jardin are open, turn it on the south side to rejoin the approach by climbing north between the Jardin and the left side of the eastern spur (2½hrs).

Ascent Cross the bergschrund (or take the rocks on the left bank). Climb the couloir (180m 45°) to come out on the snow or mixed crest. Leave this fairly soon to cross to the south-east side to reach, by going up right, the foot of the rock step at around 3700m (1½hrs).

Go up right on ramps then come back left to near the crest in two or three zigzags (cracks/chimneys, 4b/c, verglas) to reach the summit of rock step around 3800m (1hr). Move diagonally left up the steep snow slope to reach the fine ridge leading to the foot of a rock tower. Turn it on the right and climb quite a steep, open dièdre (mixed, 30m). Go to the left of a final block and come out at the summit. 40mins, 3–4hrs from the bergschrund.

Descent on the south-east side
From under the summit an abseil regains the foot of the rock tower. Then make five or six abseils in a couloir heading down south-east, keeping mainly on the left bank, then well to the left to end on the rock step which marks the finish (1½hrs, abseil anchors may require back up).

On reaching the snow, go diagonally left towards the fall-line from the Col des Droites. Then, keeping right, descend the whole of the east slope (45° then 40°) to the bergschrund (c.3230m, 45mins). Make a wide turn from left to right (crevasses), to reach the Talèfre Glacier and the track coming back from the Courtes. Pass to the south of the Jardin and pick up the path to the hut on the south-west (45mins, 3hrs from the summit).

Les Droites: the snowslope below the final ridge of the eastern spur.

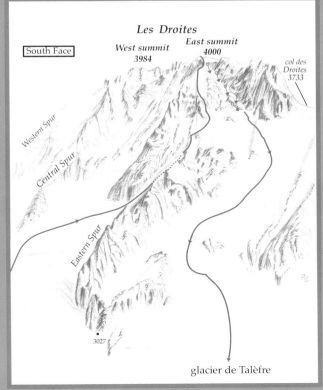

Les Droites

South Face

West summit
3984

East summit
4000

col des
Droites
3733

Western Spur

Central Spur

Eastern Spur

3027

glacier de Talèfre

Looking across the
Talèfre Glacier to Aiguille
Verte (centre right) and
Les Droites (far right) —
its eastern spur picked
out by the sun and the
descent couloir and
slopes largely in shadow
(farther right).

Note (for UK/US edition)
On descent, if the
avalanche risk on the Col
des Droites slopes is too
high, one can either
return down the ascent
route or, from the end of
the abseils down the
south-east couloir,
traverse south-west to
regain the crest of the
eastern spur and abseil
down the right bank of
the approach couloir.

TRAVERSE

Traverse of Les Courtes
Start Couvercle Hut (2687m) 3hrs from Montenvers. Approach p94.
Difficulty III /PD+ to AD according to conditions. An entirely mixed route on an airy ridge.
Time Hut to summit 4–5hrs, traverse and descent 3hrs to the bergschrund, then 1hr to the hut.
Height gain 1170m.
Conditions June – September, in stable snow conditions after a good overnight frost (to stabilize avalanche-prone slopes). More pleasant in the descent when there is a good snow cover.
Equipment Crampons, ice axe, rope for 30m abseil, 2 ice pegs, slings, karabiners, nuts.
First ascent *WNW Ridge:* O. Schuster and A. Swaine, 17 August 1897. *SE Ridge:* E. Fontaine with J. Ravanal and L. Tournier, 11 August 1904, in descent.

Reaching the Talèfre cirque involves a long approach up the Mer de Glace, an arduous moraine, and finally the endless iron ladders of Les Egralets. One eventually arrives at the Couvercle huts – the old one with its incredible granite lid and the new with a fine view of the Grandes Jorasses. Traversing Les Courtes is not particularly difficult but does require a good grasp of basic techniques and a high level of fitness. Most of all, good route-finding is required on some sections such as finding the letter-box on Aiguille Chenavier, taking the correct line around the Aiguille Qui Remue, and locating the ramp system that is the key to the bergschrund-crossing on the descent. Aim to be on the summit ridge at daybreak when the snow will be firm for efficient cramponing enabling you to establish a good margin of safety. Traversing Les Courtes demands commitment: there are no quick escapes as the depths plunge on either side. There is a great sense of remoteness and it is obvious that matters could become serious in the event of bad weather. Back at the hut after a safe descent you will bathe in the joy of existence, enjoy a beer with friends ... and certainly feel like more of a mountaineer.

Arriving at the summit of Les Courtes, with the upper slopes of Route 7 on Les Droites behind.

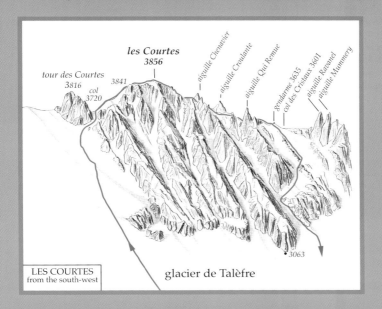

les Courtes
3856

tour des Courtes
3816
col
3720
3841

aiguille Chenavier
aiguille Croulante
aiguille Qui Remue
gendarme 3635
col des Cristaux 3601
aiguille Ravanel
aiguille Mummery

• 3063

LES COURTES
from the south-west

glacier de Talèfre

Starting up the Égralets ladders.

Ascent by the WNW Ridge
From the hut, follow the path north-west (cairns), which leads to the Talèfre Glacier (30mins). Pass south of the Jardin and around 2850m head north diagonally across the slopes leading to the Col des Droites basin.
At about 3350m, make a rising traverse right up a snow slope for a short distance and before the top, traverse right under the Tour des Courtes, and climb the gully leading to the col of the same name (3720m, 3½hrs). Above, after a steep slope, follow the airy snow ridge to the summit. 1hr, 4–5hrs from the hut. Diagram p33.

Traverse and descent
Descend the ridge crest south-west to the exit of the North-East slope (large boulders or snow). This leads to the Aiguille Chenavier. A letter-box can be seen to the right of and a little below the summit (boulder with slings). Reach it by crossing a slope for around 30m (poor rock, snow in early season). Go through it to reach the south-east slope. (If the letter-box is blocked with snow, go to the summit of the Aiguille Chenavier and reach the south-east slope by a 30m abseil.)

Continue descending near the ridge crest, then traverse right, to the foot of the Aiguille Croulante (mixed or boulders). Reach a prominent col a little before the Aiguille Qui Remue (snow, view down the north slope of the Col des Cristaux). Turn this point on the Argentière (east) side by the snow slope or steep mixed ground (50m). Arrive at a col (many crystals). From there, follow the widening ridge, to an obvious small gendarme (2hrs from the summit).

On the summit ridge of Les Courtes with Mont Dolent just to its left.

The small gendarme marks the snow or scree slope of the descent on the Talèfre side. Where it steepens farther down keep progressively right (cairns, ledge system with snow patches or gullies to cross) to reach the side of the couloir descending from the Aiguille Croulante (steep wall on the right). Go down its left bank to the bergschrund and cross this, keeping left, to reach the Talèfre Glacier (1hr). Descend in a south-easterly direction to the hut. 4hrs from the summit.

NORMAL ROUTE

Start Couvercle Hut (2687m); 3hrs from Montenvers. Approach p94.
Difficulty I/PD+ Crevassed glacier terrain, with two short rock steps and an airy ridge (35°– 40°).
Time 4–5hrs. Descent 3hrs.
Height gain 1074m.
Conditions Good, stable snow is essential. Harder in late season if crevasses are too open.
Equipment Crampons, ice axe, rope, 2 ice pegs selection of slings, 2 quick draws, 4 karabiners.
First ascent Miss Isabella Straton with J.E. and P. Charlet, in 1875.

Isabella is a romantic first name quite worthy of appearing on a map, but who is the mysterious person for whom the peak was named? The young British aristocrat Mary Isabella Straton soon left behind her mansions and afternoon teas to make her impression in the man's world of the alpine peaks. She is one of several female alpine pioneers who (with their guides) shamelessly scoured Mont Blanc massif in the latter years the 19th Century. Not content with snatching a clutch of prestigious ascents accompanied by her favourite guide, Jean Esterel Charlet, she ended up marrying him. Their joint exploits included the Moine, the first winter ascent of Mont Blanc, the Persévérance, and, deep in the Talèfre basin, the first ascent of a previously overlooked peak that was then suitably titled Isabella.
The route is mainly on glacier terrain – its mild technical difficulty permits its charms to be fully enjoyed on the gradual ascent during the last hours of the night. There are remarkable sections beside an icefall, daring zig-zags through deep crevasses, beautiful ridges shaping the pure line of the route leading to the edge of the material world, where the horizons are vast.
This climb is effectively the descent route after an ascent of the North Face the Triolet. An ascent is thus of value if that great climb features in your future plans.

Normal Route by the West Spur
From the Couvercle Hut, descend eastward by the path to reach the Talèfre Glacier. Pass below the Jardin and go up the moderately inclined slopes more or less north-east, describing a large arc from left to right. So reach a glacier ramp (the Courtes Glacier) which runs below the south side of the Arête des Rochassiers. Climb the right bank at first, then at around 3350m, reach a little cirque at the foot of the séracs coming out of the Trient Plateau. Then traverse due south, to the base of a shallow-angled snow are just below the big rocky spur at the right extremity of the face. (Pte.3401m). Follow this ridge – the West Spur (25° for about 100m). Climb the short, easy rock step immediately above. This leads to a second snow ridge, sharper and very aesthetic, about 150m in length (30° then 35°). Follow it (impressive view of the séracs on the west face, on the left). After a short traverse, follow a shoulder then a rib going eastward. This leads to the start of the final rock tower above the Col du Triolet. From there, cross the bergschrund and go up right following the north-east ridge (snow, then rock flakes and blocks, 3/3c) to the North summit, the highest point. 4–5hrs. Diagram p33.

Descent By the same route, 2–3hrs.

The shoulder of Pointe Isabella's West Spur with Mont Blanc in the background.

INTE ISABELLA
alèfre (west) aspect

aiguille
du Triolet
3870

pointe
Isabella
3761

col du Triolet
3703

arête des Rochassiers

plateau de Triolet

pointe
3401

glacier des Courtes

Aig. du Triolet and the Trient Plateau.

SOUTH-WEST COULOIR AND TRAVERSE

Start Leschaux Hut (2450m). Approach p95.
Difficulty II/AD–. A snow couloir at 45°–50°. Descent: Couloir at 35°–40° for 150m.
Time Approach 3hrs. Ascent 2hrs. Descent 2hrs.
Height gain From hut to summit 1200m, including 400m in the couloir.
Conditions You need stable snow (there is a huge avalanche cone). Springtime preferable.
Equipment Crampons, ice-axe, one extra tool per party (for the bergschrund), rope, 3 ice pegs, deadman, 3 quickdraws, slings, nuts, helmet, headlamp.
First ascent C.E. Matthews, A. Adams-Reilly, M.A. Ducroz, M. Balmat, 7 July 1866.

The Aiguille de l' Éboulement is the most inappropriately named peak in the whole range and that it should be re-named, if not the Jewel in the Lotus Flower, then at least the Aiguille of the Far Horizons! The name of a peak is of crucial importance. In this case, an enormous rock slide was the reason for the original name and you need to forget that in order to treat yourself to this route. It has no particular difficulty, but is in one of the most secret parts of the massif.

The natural features in which this jewel is set are prestigious, intimidating too: the Grandes and Petites Jorasses and the Aiguille de Leschaux. But in the course of the climb these peaks will become familiar, seen to advantage from an interesting viewpoint

If you wish to consider the traverse in advance from a distance, the best viewpoint is from the Périades from where you can also see the elegant snow couloir leading to the summit.

Technically, only the bergschrund is difficult

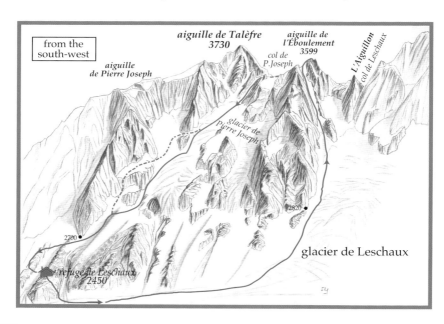

L'ÉBOULEMENT (3599м)

The exit from the snow couloir with the
Grandes Jorasses in the background.

The eastern branch of the Leschaux Glacier
that is the approach to the South-West Couloir
of the Aiguille de l'Éboulemont, clearly seen,
sunlit, in the centre of the picture.

North Face

Grandes Jorasses

aiguille
de Leschaux

Petites
Jorasses

col des
Hirondelles

aig. de
l'Éboulement

glacier de Pierre-Joseph

refuge
de Leschaux

glacier de Talèfre

glacier de Leschaux

Mer de Glace

The Aiguille de l'Éboulement and the Pierre-Joseph Glacier. Behind on the left, the Grand Combin and the Matterhorn. To the

being vertical for a few metres and calling for a little ice-craft. The angle is comparable to the Table Couloir on the Aiguille du Tour, but the descent is rather longer and on a glacier where there is less chance of finding a track.

The short descent couloir has a steep section at the top. This gives an excellent test of balance and will get you used to exposure. If unsure, climb down it facing, in but if the snow is firm and the crampons gripping well, try it facing out.

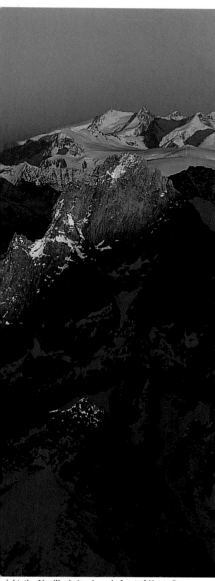

right, the Aiguille de Leschaux in front of Monte Rosa.

Other than the aforementioned neighbouring peaks the new aspects of the Grand Combin and the distant Valais peaks are another visual reward to treasure.

South East Couloir and Traverse

Pre-inspection The evening before the climb it is useful to inspect the return route (above the hut near the water supply), as it is not easy to pick out when coming from higher up.

Approach Go down to the glacier (15mins) and climb south-east on its right bank. Around 2700m, go east on to its upper branch. Pass a rock island at about 2800m and go north-east up sustained, crevassed slopes, keeping to the right bank. Reach the bergschrund of the couloir at 3150m just to the left the South-West Ridge of the Aiguille de l'Éboulement. 3hrs, diagrams pp42–43.

Ascent If the bergschrund is uncrossable, climb the rocks on the right bank (50m, nuts, slings) and then move back into the couloir. Go up it, keeping right as you near the end to come out at a col. Climb a short rocky ridge (south) to the summit. 1½–2hrs.

Descent by the Pierre Joseph Glacier

Come back to the col and descend an easy ridge and snow slope (north, then north-west) to the Col de Pierre Joseph (3500m, 10mins). From there, descend a little, then go up to the north, passing to the right of a split gendarme to enter the couloir which opens on to the west side. After 50m, go north along a snowy ledge, first climbing, then descending. This brings you to the descent couloir (35°–40°, 100m), leading to the glacier (bergschrund around 3300m). The couloir to the left of the gendarme is also feasible, but has poor rock at the top, then quite steep snow or ice (40°–45°). Make a diagonal descent towards the right bank of the glacier, which is followed for its whole length. At about 2750m go diagonally right to meet the old lateral moraine. Directly below point 2700m (bottom of the SW Ridge of the Aiguille Pierre-Joseph), go straight down to grass slopes above the rock steps overlooking the hut. Go diagonally right (north-west, intermittent path) and reach and follow a little stream before coming back right to the hut. 2hrs.

Note If the glacier is too open, from the upper plateau bear right to reach a snow or earth saddle below the SW Ridge of the Petite Aiguille de Talèfre. By easy slabs and a scree (or snow) gully, reach the top of the lateral moraine.

Start Torino Hut (3371m). Access by cable car on the Italian side, or on the French side by crossing the Vallée Blanche (2½hrs or gondola). See p95.
Difficulty III/AD High altitude, corniced snow ridges. Mixed on approach and summit.
Time 4hrs from the hut. Return by the same route, 3hrs.
Height gain 630m.
Conditions June – October. Check that there is not too much snow.
Equipment Crampons, ice axe, rope, slings, 2 ice pegs, 3 quickdraws, 4 karabiners.
First ascent E. Allegra with L. Croux, P. Dayné and A. Brocherel, 18 July 1900.

The Rochefort Ridge, the natural frontier between France and Italy, is one of the most popular high mountain excursions in the Alps. Its sustained altitude at around 4000m should be taken into account as it is essential to get acclimatised first so as to be able to enjoy the surroundings without physical distress. The difficulties are not great; the ease of access and the rapid out-and-back to the foot of the Dent du Géant make it a fine excursion for competent parties moving quickly.

Photographers will revel in the tracery of its cornices – fine snow sculpture etched by the morning sun as one advances along the twists and turns of the ridge. In the event of a slip, the classic tactic of jumping over the other side will need to be employed, a possible drama that concentrates the mind. The rock buttress leading to the Aiguille de Rochefort will show you whether you can choose the logical line, as well as how confident you are when down-climbing the rock, rather than doing it in two abseils, if you are coming back on the Torino side. From the broken rocks on top of this peak, the views to the Jorasses, and the from Aiguille Noire to Mont Blanc and then to the intricate Mont Blanc du Tacul are exceptional.

Looking west along the Rochefort Ridge to Mont Blanc.

The view to the east along the ridge.

Dent du
Géant
4013

Salle à
Manger

aiguille
de Rochefort
4001

gendarme
3665

3933

col de
Rochefort

aiguilles
Marbrées

3516

glacier du
Géant

ARÊTE DE
ROCHEFORT
from the south-west

From high in the Vallée Blanche one can enjoy this panorama from the Drus to the Dent du Géant.

The Rochefort Ridge

Approach From the Torino hut cross the Géant Glacier to the north-east, a little north of the Col du Géant, and pass the foot of the Aiguilles Marbrées. Go to the top of the glacier to the left foot of the couloir, which splits the bottom of the rocks, more or less directly below the Dent du Géant. Cross the bergschrund (around 3600m) and climb the couloir to come out on a shoulder to the right of a reddish gendarme at 3665m. (If the couloir is too icy, zig-zag up a system of easy ledges on the edge of its right bank.)

Go up right on broken rock, sometimes with snow, climb a cracked dièdre (3b) and continue bearing right, towards the ridge which bounds the face on the north-east. Go along its base leftwards (little easy walls) to the foot of a yellow monolith which is turned by the right, descending a little at first (Italian side, recent rockfall, move quickly!). A short easy slope slants up right to the Salle à Manger, (breakfast spot) at the foot of the Dent du Géant. 2½hrs.

Traverse Follow the ridge to the east turning on its right Pte.3933 if it is too sharp. Descend on the north side from the col which follows (20m, steep, exposed) using a back-up piton for belay or abseil according to conditions.

Continue eastward along the ridge, turn a few gendarmes mainly by the right, and reach the foot of the rocks leading up to the Aiguille de Rochefort. Climb these, first straight up for 30m (belay), then move left into the gully which goes up to the right for 50m to the ridge, just by the summit. 1½hrs, 4 hours from hut.

Descent By the same route. A traverse taking the Mont Mallet Glacier for descent has become far more difficult since the heatwave of 2003. The crevasses have widened and it can become uncrossable! If this route is planned it is therefore important to consult the Office de Haute Montagne in Chamonix in advance.

The Aiguille du Géant at sunrise with the Rochefort Ridge silhouetted on its right.

NORMAL ROUTE
NORTH FACE
CRAMPON FUTÉ VARIANT
GERVASUTTI COULOIR

Start Torino Hut (3771m). Access by cable-car on the Italian side, or on the French side by crossing the Vallée Blanche from the Midi cable-car station (2½hrs or gondola). See p95.

Easy access with the perfect viewpoint in the heart of the range, the Tour Ronde might still become the prime example of peaks being abandoned due to climate change. Its moderate altitude makes it the ideal target for the warming which is affecting the snow resources of the massif.

The scars of the 2003 heatwave persist and the poor annual snowfalls have not been sufficient to make up the deficit. Glaciers open, the mountain-sides lose their permafrost, turning their rocks into an unstable kind of conglomerate. Thus the now denuded Normal Route possesses numerous loose boulders ready to fall, either spontaneously or dislodged by climbing parties.

Those who are not familiar with the alpine ranges and who have not followed these changes, can be greatly at risk from following certain routes blindly. Before attempting them consult the guides' offices and mountaineering-related websites (see p96). The Tour Ronde is now only sensibly tackled after a good

snowfall which has become well stabilised. Otherwise, *don't go* or, if you must climb this peak, which is still a very strategic viewpoint, start from the Col d'Entrèves and keep below the crest of the ridge on the Brenva side – a route requiring route-finding competence and a good level of fitness. Since the advent of the technical axe, the North Face has come to be treated as a training ground with some parties even failing to make a night-time start. The central runnel tends to dry up and channel all kinds of debris coming from above. The Crampon Futé variant keeps right of the classic route and (in suitable condition) avoids some of the risks. The west-facing Gervasutti Couloir holds more snow and offers a regular slope embellished by a finish by the North Face, but its bergschrund soon opens and can present a formidable obstacle. Until and unless global cooling restores the natural order, you will have to seize a suitable moment to reach the summit.

The final slopes of the Normal Route of the Tour Ronde.

Approaching Combe Maudite, with Tour Ronde on left, Mont Maudit (centre) and the spurs of Mont Blanc du Tacul (right).

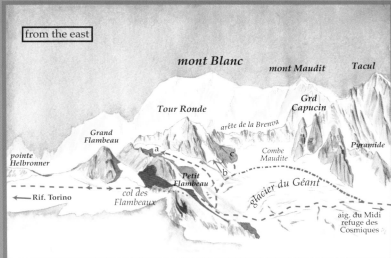

from the east

mont Blanc

mont Maudit

Tacul

Tour Ronde

Grd Capucin

arête de la Brenva

Grand Flambeau

Pyramide

Combe Maudite

pointe Helbronner

a

c

b

Petit Flambeau

← Rif. Torino

col des Flambeaux

glacier du Géant

aig. du Midi refuge des Cosmiques

(a) Normal Route: East Slope and South-East Ridge

Difficulty II/PD mixed : Snow slope and ridge at 40°, classic crampon work, easy rock.
From the Col d'Entrèves (alternative) II/AD Ice and mixed, route not as obvious as it seems.
Time Hut to summit 3–4hrs. Descent 1–3hrs.   **Height gain** 421m.
Conditions If the slope is avalanche-prone or too bare, take the South-East Ridge from the col.
Equipment Crampons, axe, rope (possible 50m abseil above the Col d' Entrèves), 2 ice pegs, 3 slings, 3 karabiners.
First ascent J.H. Backhouse, T.H. Carson, D.W. Freshfield, C.C. Tucker with D. Balleys and M. Payot, 22 July 1867.

Approach From the Torino Hut, head north-west over the Col des Flambeaux (3407m), then descend diagonally north-west below the Col Orient and Aiguille de Toule to a flattening around 3300m. Head west then south, rising diagonally, to go up the glacier valley coming out of the Col d' Entrèves. Gain the slope which starts immediately left of the wide rock buttress (about 250m left of a direct line down from the summit). 1hr.

Ascent East Slope and South-East Ridge. If there is enough snow and conditions are stable, cross the bergschrund and go up the slope. After about 150m traverse right under a block, to join the South-East Ridge after a left-right S-movement. Follow it (north) passing a few boulders. The ridge turns into a short snow slope; climb this to the foot of the summit rocks, which are climbed from left to right, to reach the Virgin statue. 1½hrs.

Alternative If there is a lack of snow on the East slope, continue up the glacier valley to a snow col just to the right of and below the Col d' Entrèves, from which it is separated by a little outcrop of rocks (around 3540m). Climb the South-East Ridge from the col in 4 pitches of 45m (2 on ice at 50° and 2 on easier mixed ground). Climb parallel to and right of the rock ridge which stands above the col to the west. (Note: On the second pitch, keep well left and avoid taking a ledge going up right as that leads to a difficult dièdre which, although it gains the ridge, complicates access to the Brenva side.)
After a little granite spire and a shallow-angled slab, a gap on the SE Ridge is reached, just left of a 10m yellow wall (the finish of the difficult dièdre, 40mins). From the gap, swing over to the Brenva side, which is traversed north-west, by ledges below the ridge, the level which varies with the terrain (scree and

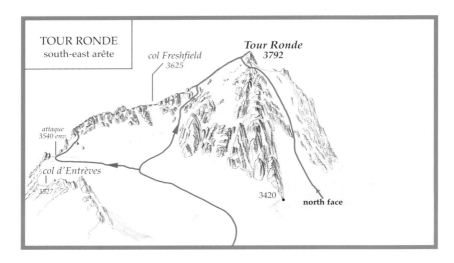

TOUR RONDE
south-east arête

col Freshfield
3625

Tour Ronde
3792

attaque
3540 env.

col d'Entrèves
3527

3420

north face

At the exit from the Gervasutti Couloir of Tour Ronde, with the south-east side of Mont Blanc du Tacul behind.

earth often with snow or ice, a few cairns). After about an hour, go up right to rejoin the crest of the ridge, beyond some looming gendarmes, visiting if desired, Col Freshfield. After turning a little monolith, still on the Brenva side, reach the level junction with the East Face. From there, gain the summit by mixed ground, a steep snow rib, and then the short final step. 2½–3hrs from the col.

Descent By one or other of these two routes, depending on conditions and the time of day and numbers and positions of other climbers. 1–3hrs.

(b) North Face by the Classic or the Crampon Fûté variant
Difficulty *Classic* II/D– A short ice route, with one pitch at 55°– 60°, the rest between 40° and 50°. *Crampon Fûté* II/D, 55°– 60°, more exposed.
Time Approach: 1hr from the Torino, 2hrs from the Cosmiques. Ascent: 2–3hrs. Descent: 1–3hrs.
Height gain 350m on the face.
Conditions Best with a stabilised snow cover, May – September, sometimes in winter. In a dry summer, the face soon turns to ice and even underlying rock presents objective dangers.
Equipment Crampons, technical axes, rope (50m abseil sometimes needed on the descent to the Col. d' Entrèves), 6 ice pegs/screws, selection of slings, 4 quick draws, 4 karabiners, a few nuts.
First ascent Alexis Berthod and F. Gonella, 23 August 1886.

Tour Ronde's North Face in acceptable condition.

Approach From the Torino Hut, head north-west over the Col des Flambeaux (3407m), then descend north-west below Col Orient and Aig. de Toule. On easier ground (c.3300m) traverse west below the NE Spur of Tour Ronde to Combe Maudite and then south-west up to the foot of the North Face.

North Face
Classic: Cross the bergschrund and climb straight up to the narrows (around 130m at 50°). Climb the gully directly for two pitches (about 80m, 55°– 60° depending on how much ice, sometimes a few rocks). Climb the final slope that broadens out and becomes progressively less steep (45° then 40°, 150m). On the left there is a gendarme lying down. The slope runs out against the summit rocks. After passing the gendarme, cross a couloir to the left and get on to the east side, thereby turning the summit rocks (big blocks, a move of 3b, 150m). A short snow slope leads to the SE Ridge, which leads to the foot of the last rock pitch. Easy cracks and a traverse right bring one out near the summit statue. 2hrs.

Alternative
Crampon Fûté Variant: Follow the Classic for two or three pitches, then go rightwards up a vague steep couloir on mixed ground to rejoin the Classic, where it comes out to the left of the summit rocks. 3hrs.

Descent By the Normal Route, p52.

Gervasutti Couloir
Difficulty II/AD, uniform slope at 50°
Time Approach: 1½hrs. Climb: 1½hrs. Descent: 1–3hrs.
Height gain 350m.
Conditions Stable snow. Check on the condition of the bergschrund, which can be uncrossable.
Equipment As for the North Face except that 4 ice screws (per party) should suffice.
First ascent R. Chabod and G. Gervasutti, 27 July 1934.

Approach From the Torino Hut, follow the approach for the North Face. Go past the start of that route and move round the West Pillar. Climb the basin that opens south to reach the obvious couloir splitting the whole West Face. 1½hrs. Colour sketch p51.

Ascent Cross the bergschrund and traverse the gully to the left bank, which is climbed with a finish up a short runnel on mixed ground to reach the the snowy shoulder above the North Face (1hr). From there, reach the summit, turning the summit rocks on the left.
Descent By the Normal Route, p52.

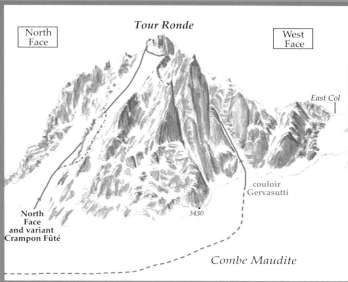

Tour Ronde

North Face

West Face

East Col

couloir Gervasutti

3430

North Face and variant Crampon Fûté

Combe Maudite

The view across Combe Maudite from below the Col du Diable on Mont Blanc du Tacul. La Tour Ronde is on the left and the three cols on the right are Ronde, Trident, and Fourche.

MIDI-PLAN TRAVERSE

Start The Aiguille du Midi cable car station (3795m) or Cosmiques Hut (3613m). Approach pp 94, 95.
Difficulty III/AD+ Exposed, high-altitude snow ridges (cornices) and mixed ground (rock sections of 3), plus a crevassed glacier route (quite hazardous after midday) if the descent to the Requin Hut is taken. [The climb was traditionally graded PD+, and the new grade reflects barer (sometimes icy) snow slopes, greater difficulties on the ascent of the Plan and harder crevasse conditions on the Envers du Plan Glacier. *Editor*]
Time Traverse: 4hrs. Return to Midi: 4hrs. Descent to Montenvers: 3hrs.
Height gain 200m of ascent (mainly on the Plan, plus 250m if starting from the Cosmiques) on a ridge with a generally descending line. 2000m of descent from the Midi to Montenvers.
Conditions The ridge should be neither too dry nor too snowed-up, and well frozen.
Equipment Crampons, ice axe, rope for 30m abseils, descender, selection of slings, 2 ice pegs, 3 quick draws, 4 karabiners, warm clothing.
First ascent G.W. Young with J. Knubel, 10 August 1907.

Sometimes it is possible to experience something very fine without having to cross oceans or juggle timezones. The landscapes being travelled need to offer contrasts, with technical difficulties that concentrate the mind enough to ensure a

At the start of the Traverse.

satisfied tiredness at the end of the day. The Midi-Plan Traverse is ideal in this respect. Here is a wonderfully exposed climb, with frequent drops and rises and icy walls sweeping down toward Chamonix on its north side and vaste snowfields stretching into the distance on its other flank.

The ridge appears level from the valley and whether one is a walker or climber, it *seems* feasible. The proximity of the cable car reinforces this, reality proves quite otherwise. Some delicate and distinctive places will be encountered: a very sharp, steeply descending ridge at the start, then a section of mixed, with some exposed traverses. More snow domes follow, then a tricky descent on good rock and finally some rock climbing to reach the superb summit of the Plan, with its magnificent view across the Chamonix Aiguilles.

At the start of the season, the descent by the wildly situated Envers du Plan Glacier will take you deep into the bowels of the range, where the ice snakes lurk among unfathomable depths. Then appears a haven of safety and comfort — the Requin Hut strategically placed above the great Tacul and Leschaux glaciers. The journey ends at Montenvers, and its interesting cog railway that transports you leisurely back to Chamonix.

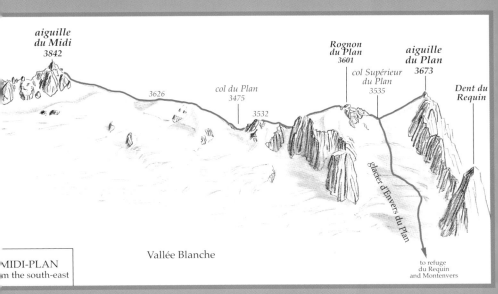

aiguille
du Midi
3842

3626

col du Plan
3475

3532

Rognon
du Plan
3601

col Supérieur
du Plan
3535

aiguille
du Plan
3673

Dent du
Requin

glacier d'Envers du Plan

Vallée Blanche

to refuge
du Requin
and Montenvers

MIDI-PLAN
m the south-east

Autumn light on a backdrop of the Verte, the Droites and the Courtes. Aiguille du Plan is on the left.

The Midi-Plan Traverse

From the galleries on the Aiguille du Midi, go down the airy North-East Ridge and then follow the fine snow crest, which becomes narrower (cornices) and ends with a steep descent to the Col du Plan (3475m). Continue along the ridge, turning rocks on rognon 3532m by an exposed traverse on the north side and thence to a snow saddle. Again on the Chamonix side, go up the fairly wide snow or scree couloir which leads, after an easement, to the Rognon du Plan. Follow the crest, and then (on the east side) descend a steep system of chimneys, ledges and cracks on rock or mixed ground (two 30m abseils possible). By a delicate traverse, gain the Col Supérieur du Plan (3535m, 3hrs). Climb a short slope to reach a snow saddle at the foot of the rock step below the summit of the Aig. du Plan and ascend this by a turning movement on the west side to reach the top, a fine, airy flat slab (3b, 1hr).

Descent Return to the Aiguille du Midi by the same route, but if the Envers du Plan Glacier is not too open, descending to Montenvers may be of interest. Return to the Col Supérieur du Plan and from there descend a steep slope to reach the Envers du Plan Glacier. Go down it, choosing the best route through wide crevasses and some séracs, in the middle at first, then by an 'S' movement on the left bank. At around 2900m, turn left below the buttresses of the Aiguille du Requin, to come back right (rock step, chains) and reach the moraine which leads to the Requin Hut (1½–2hrs from the Plan).

From below the hut (2516m), following a path N then NE to reach ladders that go down to the Mer de Glace. Descend this on its left bank, hugging the lower slabs of the Envers (rock-climbing area). When the glacier heads north-west, go into the middle then trend steadily toward the left bank

Early sunlight on the Midi-Plan Traverse.

(look for the big white rectangle painted on the rock) and quit the glacier on ice ridges. At about 1750m start climbing ladders, then by a good path, go up northwards to the Montenvers station (1913m), 1½hrs – 2hrs from the Requin Hut.

AIGUILLE DU MID

COSMIQUES ARÊTE

Start Aiguille du Midi cable-car station (3795m). See p95.
Difficulty II/AD mixed, with a rock pitch (smooth crack 4b) and some complex ropework.
Time Approach 45mins. Ascent 2–3hrs.
Height gain 268m from the Col du Midi.
Conditions Snow should be re-frozen, and stays firm longer as the ridge is west-facing.
Equipment Crampons, ice axe, rope for 20m abseils, descender, selection of slings,
3 quick draws, 4 karabiners, nuts.
First ascent G. and M. Finch, 2 August 1911.

This varied mixed route takes its name from the old laboratory (study of cosmic rays) on the Col du Midi which was replaced by the comfortable Cosmiques Hut.
On the approach you will pass below the South Face of the Aiguille du Midi, with its famous and impressive Yosemite-like wall climbed by the celebrated Rébuffat/Baquet Route (1956).
By contrast the Cosmiques Ridge presents a menu of classical difficulties: mixed ground, steep gullies, smooth slabs, rope manoeuvres, narrow ridges, etc. It lacks nothing except length. You even get a public reception committee when, to the clicking of many cameras, you emerge on to South Peak's terrace, and the descent is a mere few steps down to the cable car. Such diversity in one route makes it an ideal training climb for those with bigger aspirations. However, if this is your first visit, or even if making a repeat, or if you are just enjoying a pleasant half-day's outing, remember that the climb is at high altitude, west-facing, and very exposed to the weather. The Midi is the first to encounter the westerly squalls that can transform an amiable ascent into a tense struggle. So stay alert for changes in wind direction, and take some warm clothing.
That apart, this is a fine climb in a grandiose setting. The rock is protogine, 40 million years old, and cherished by alpinists for its tawny colour and velvety grain, smoothed by aeons of battering by glacier, lightning and storm.

Route From the cable-car station galleries, descend the North-East Ridge (exposed, care required), then turn down southward below the rocks of the South Face. After passing another prominent rock face (the Cosmiques Spur) make a short ascent until level with the Cosmiques Hut on your left. Go west to reach a level section (3593m) and gain the foot of the ridge above the Abri Simond (hut cableway). 45mins.

Ascent Start to the right of the ridge crest (obvious flake) and climb the couloir (snow in early season, mixed or pure rock later), to a gendarme (3731m). Follow the ridge (snow then rock) and turn a first large tower on its right (20m abseil). From its base, either:
(a) go back to the ridge and turn a second tower on its left 3b, large blocks followed by a little descent gully (possible 20m abseil if icy) that leads on to the now level ridge. Or,
(b) if the snow is stable and frozen, the second tower may be turned on the right, going diagonally up the steep slopes leading back to the ridge.
By the snowy crest, reach the foot of the final rock step (orange slab). Climb the smooth crack (7m, 4b, metal wedge) and at the top, follow a little ledge to the right and by a flake, come back left onto a platform above the orange slab. Then transfer to the west side to climb the exit chimneys (sometimes snow or ice) to emerge on the snow shoulder below the South Peak's terrace (ladder). 2–3hrs.

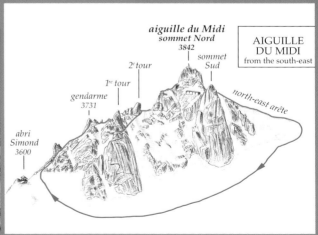

aiguille du Midi
sommet Nord
3842

AIGUILLE
DU MIDI
from the south-east

sommet
Sud

2ᵉ tour

1ʳᵉ tour

north-east arête

gendarme
3731

abri
Simond
3600

At the finish of the Cosmiques Ridge.

NORMAL ROUTE
CONTAMINE-GRISOLLE
CHÈRÉ COULOIR

Start The Aiguille du Midi cable car station (3795m) or the Cosmiques Hut (3613m).
Approach pp94–95.

With its easy access by cable car, Mont Blanc du Tacul's North-West Face is useful to tune up glacier skills, for training prior to trying one of the couloirs or pillars on its North-East Face, for acclimatising, or just for sightseeing.

Before starting you should check on the conditions on the mountain as these vary considerably. As with other places in the massif, heat and lack of snow change things, but here the results can be dramatic. The

climb, traditionally graded PD, can be as hard as AD+ in a lean season. On the right (north-west) two wide crevasses, soon after the start, can become difficult, and if big avalanches from the cwm above are not filling them regularly, they soon become impassable.

Another result is that these glacial slopes when stripped of snow expose huge séracs, and these threaten the route.

It has thus become essential to get up-to-date information before undertaking *any* route on the mountain. While it is easy to turn back on the Normal Route if circumstances dictate, it is important to be aware of the state of the North-West Face if reaching the summit by another route, or if doing the traverse from Mont Blanc and Mont Maudit and hoping for an unproblematic descent.

Assuming the climb is in its normal state, you will be impressed by the unfolding vistas of high snowfields, as well as the huge bulk of Mont Blanc itself. You will also soon discover the true state of your personal acclimatisation.

Those climbing at a higher standard can approach by one of the short steep routes that take the small triangular North Face. The Contamine/Grisolle is the most amenable of these – an interesting mixed climb following the line of least resistance on the left side of the face to gain the snowfield below the summit rocks. These present the major difficulty and are climbed by a gully where your ice tools will show their worth. All the routes on this northern gable come out at point 3970m from where you reach the summit of the Tacul by its attractive North-East Ridge.

The Chèré Couloir provides good practice in the use of ice tools on six sustained pitches with belays in place and an exit on mixed ground among fine ruddy rocks. You will find Friends and nuts useful for protection on the sides of the gully and ice pegs and screws can also be used to advantage. The commitment is not great, as an abseil descent can be made using the insitu belays if the weather turns.

The ease of access does not lessen the inherent seriousness of the peak. Settled weather is important as the onset of a storm is a serious matter on this face, exposed as it is to fronts approaching from the west – tracks can disappear in a few minutes and after heavy snowfall, windslab is common. So stay alert and be prepared to turn back in good time.

Evening sun on the North-West Face of Mont Blanc du Tacul. The rocky North Face triangle is on the left and Mont Blanc on the upper right.

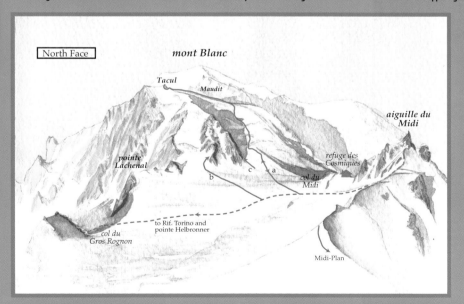

North Face

mont Blanc

Tacul

Maudit

aiguille du Midi

pointe Lachenal

refuge des Cosmiques

b

c

a

col du Midi

col du Gros Rognon

to Rif. Torino and pointe Helbronner

Midi-Plan

(a) Normal Route (North-West Face)

Start   Aiguille du Midi cable-car station (3795m). See p95.
Difficulty   II/PD—AD   Steep snow slopes, avalanche risk, crevasse & altitude problems. Often crowded.
Time   From the Aiguille du Midi to the summit: 3hrs. Return: 2hrs.
Height gain   716m, and 268m on the way back to the Aiguille du Midi.
Conditions   Keep alert for thinly covered crevasses; windslab risk after heavy snowfall.
Equipment   Crampons, ice axe, rope, 4 ice pegs, slings, 3 karabiners, warm clothing.
First ascent   Rev. Charles Hudson deviating from a 7-man guideless Mont Blanc probe. 8 August 1855.

Ascent   Galleries from the cable-car station emerge on the North-East Ridge. Don crampons and descend this carefully (exposed), and where it levels out, turn south descending steeply at first and pass below the rock buttresses of the South-East Face of the Aiguille du Midi. After passing below the Cosmiques Hut gain the plateau of the Col du Midi (3542m, 45mins). Still heading south go to the start at the middle of the face. Cross the two barrier crevasses (sometimes with ladders). Above, go straight up the slope below the séracs (periodic icefalls) or (safer) if the snow is stable go right along the bottom of the hollow to gain and climb a vague rib. At the top, the two lines meet towards the right (west). After passing under some séracs, going south, come out at the shoulder at around 4050m (1½hrs) and follow the wide, rounded ridge westwards. After a little col (4200m) climb the summit rocks by a short chimney, then snow-covered rocks to emerge at the summit cross (45 mins).

Descent   Return to the Aiguille du Midi by the same route taking care when passing ascending parties where patience is often required. If in doubt belay as this is a notorious accident area (2hrs).

Morning sun on the North Face with the Normal Route in shadow on the right. The Cosmiques Hut is nearby on the right.

Dawn light slowly reveals the slopes between the Shoulder and the summit.

Passing below séracs on Mont Blanc du Tacul's Normal Route.

(b) Contamine-Grisolle

Difficulty   II/AD+, mixed with slopes of 45°–50° and a short gully pitch (55°).
Time   Approach, 1hr. To the summit, 4hrs. Return to Aiguille du Midi, 2hrs.
Height gain   Bergschrund to the top of the Triangle, 400m. From there to the top 280m.
Conditions   Stable snow is needed, especially for the descent (be alert for windslab).
Equipment   Technical axes, 50m rope, 4 ice pegs, slings, quick draws, karabiners, nuts, friends.
First ascent   G. Gren, G Grisolle, A Poulain and M. Zeigler with André Contamine. 4 July 1968.

From the cable car descend the North-East Ridge, then head south below the South Face and thence to the Col du Midi (3532m) and SE to the lower left side of theTriangle (1hr).
Ascent Climb the slope to the left of the lowest rock spur. Continue up the snow/ice slope from left to right for about 150m to a rocky defile 45° to 50°), 2 or 3 equipped belays in the rocks of the left bank. Go up the defile and come out on a shoulder. Turn a steep rock step on its right for about 60m by ice gullies (good protection). Come back left to the ridge (easier mixed climbing) for about 50m. Go diagonally up a snow slope (35° to 40°, 60m) to the bottom of the summit rocks: climb these by an ice gully (65°) and mixed ground. Above this snow slopes lead up to the summit of the Triangle (3970m) 3hrs.
From there the North-East Ridge heads up to join the Normal Route just below the summit (1hr), 5hrs from the Aiguille du Midi.
Note The exit from the Triangle can be barred by a large crevasse and become impossible. In this case, traverse west along the slopes of the North Face (crevasses, bergschrunds) to join the Normal Route at the top of its middle cwm. Beware of windslab!
Descent By the Normal Route, 2hrs.

(c) Chèré Couloir

Difficulty   II/4 The first section is an ice gully, 2 pitches at 75°–80°, equipped for abseil descent. After this mixed ground (50°–60°) to the top of the Triangle. Possible awkward crevasses at the top?
Time   Approach 30mins. Triangle and summit, 4–5hrs. Return to Aig. du Midi, 2hrs.
Height gain   Couloir 200m plus 200m mixed to the Triangle's top. 280m to summit of Tacul.
Conditions   Apart from conditions on the route, there may windslab hazard on the descent.
Equipment   Technical axes, rope for abseil, 6 ice pegs, selection of slings including long ones, 5 quick draws, 6 karabiners, full set of nuts on wire, Friends 2 to 3.
First ascent   R. Chèré, J. Tranchant, 18 Aug 1973.

Approach Follow the Normal Route to the Col du Midi. Before starting the initial slope, bear left until directly below the gully on the right flank of the Triangle. 30mins. Prepare for the climb well away from the start, which is very exposed to sérac fall!
Ascent Climb the couloir for six pitches. At its top, three mixed pitches lead to the foot of a steeper step. Turn this on the right, by three pitches on the west side, the last leading to a little shoulder on the left, under the summit rocks. From there two mixed pitches lead to the top, the summit of the Triangle being reached by a mantelshelf (3c). 3–4hrs.
Descent By abseil from belay 6 or by the Normal Route if going to the top of the Triangle (see Contamine/Grisolle).

**Starting the Central
Route on the Triangle.**

*mont Blanc
du Tacul*
4248

*Triangle
3970*

MONT BLANC
DU TACUL
N. Face (Triangle)

→ *to the shoulder*

*2 exits
possible*

goulotte Chèré

Contamine-Grisolle

Contamine-Mazeaud
(II/AD+)

**The Exit chimney of
the Contamine / Grisolle.**

AIGUILLE DU GOÛTER
TACUL-MAUDIT TRAVERSE
NORTH RIDGE OF THE DÔME

Mont Blanc holds a special place in the imagination of all alpinists. Though not particularly shapely or arresting, its high dome of remote shimmering snowfields lends it an almost Himalayan majesty and mystery that certainly befits the highest mountain in the Alps. How can you not succumb to such magical allure?

Acclimatisation, physical and technical training, good snow conditions, stable weather, maintaining a steady pace, adequate rests, and proper diet are all part of the strategy for success on Mont Blanc. If any of these elements is lacking, you may be forced into a belated retreat and in changing conditions that could become a nightmare.

Once the summit has been reached, getting back to the valley can also be a big undertaking – choosing a route through the crevasses, hurrying below the séracs of the Petit Plateau, or going back across the notorious Grand Couloir on the Goûter all demand a certain robustness of spirit.

Don't pay too much attention to those who tell you that the summit of Europe is a walk in the park. It is true that some hardy souls whip up and down it with aplomb, but to appreciate the ascent fully, you should train for it and have a dozen alpine routes behind you before attempting to realise your dream.

Paradoxically, it is alpine novices who are most attracted to this adventure. Their ascents, usually in the company of a guide, may well be the first time they have made an alpine start, setting off into the cold alpine night by the light of a headlamp – a true leap in the dark!

The main dangers are objective, mostly related to altitude, on these vast expanses of crevassed glacier terrain: fragile snow bridges, avalanche hazard and sérac falls, the quick onset of bad weather which complicates navigation, with mist and snow, immediately making route-finding very difficult.

Moreover, wind chill can make the temperature seem more than a dozen degrees colder: you need to be prepared for this and, if necessary, go back in good time to the hut always remembering that the Vallot Refuge is only an emergency shelter (and very spartan) and not a hut in the normal sense of that word.

Having said this, the ascent can go without snags, leaving you with an indelible memory and with the ambition to reach the summit again by a different route, which will be a totally new experience and another joy.

The sun will emerge as you tackle the Bosses and as you near the summit in the early morning light, think of the joy, triumph and apprehension of Jacques Balmat and Michel-Gabriel Paccard as they arrived at the top on the evening of 8 August 1786!

Aiguille du Midi and Mont Blanc du Tacul (left), Mont Maudit, Mont Blanc, Dôme du Goûter and Aiguille du Goûter (right).

Starting the Bosses Ridge.

(a) Aiguille du Goûter and the Bosses Ridge

Start   Goûter Hut (3817m). 5hrs from the Nid d'Aigle station. See pp95 and 96.

Difficulty   III/PD The hut approach is a mixed route in itself (with objective dangers) and the cumulative effect of two days' effort plus the effects of altitude, will be a factor in success or failure. There is exposed climbing on the summit ridge and accurate navigation is needed on the Dôme du Goûter in poor visibility.   **Time**   4–5hrs. from the hut. Descent 4–5hrs.

Height gain   1500m to the hut and 990m to the summit. Descent 2500m.

Conditions   Snow should be well re-frozen and stable and there should be a good weather forecast.

Equipment   Snow gear and rope (50m needed on Mont Maudit), 3 ice pegs, slings,3 karabiners, ski stick, headlamp (+ spare battery!), warm clothing, map, compass, GPS, altimeter, glacier cream, spare gloves, bivi sack.

First ascent   L. Stephen and F.F. Tuckett with M. Anderegg, J.J. Bennen and P. Perren, 18 July 1861 (though all its component parts had been climbed earlier).

Over time, this route has shown itself to be technically the easiest way up Mont Blanc, with the added advantage of having the highest starting point. Nevertheless, the long approach to the hut should not be under-estimated, with its awkward ground and high stonefall risk, particularly at the crossing of the Grand Couloir.

Despite its over-overcrowding, the ease of retreat in case of problems, the progressive difficulty of the Bosses Ridge, and the extraordinary elegance of the summit ridge, explain this route's perennial popularity.

Ascent   From just above the hut follow the horizontal ridge going south-east, and then climb the slopes of the Dôme du Goûter, passing below the summit on your right. Go on to the Col du Dôme (ill-defined, 4237m) then, after a short slope, pass the Vallot Hut (emergency shelter and radio, 4362m). From there to the summit, first climb the Bosses Ridge and then follow the narrow crest of the West Ridge that leads airily to the summit.

Descents: pp74, 75.

Starting from the Gôuter Hut.

The Bosses Ridge seen from the Vallot Hut.

North-West Face

mont Blanc

aiguille de Bionnassay

refuge du Goûter

aig. de Tricot

refuge de Tête Rousse

glacier de la Griaz

baraque des Rognes

glacier de Bionnassay

to Nid d'Aigle

(b) The Tacul-Maudit Traverse to Mont Blanc (the 'Three Monts')

Start Cosmiques Hut (3613m), reached in 20mins from the Aiguille du Midi cable car. See p94.
Difficulty III/PD + to AD depending on conditions. Slopes of 45° with sometimes tricky sections (ice, bergschrunds, crevasses), and the altitude, length and inescapability are additional risks (see p68).
Height gain 1600m, with 2500m of descent.
Conditions June – September, stable snow. Check conditions on the Tacul and the slope up to the Col du Mont Maudit and note comments in the previous chapter.
Equipment The same as that listed on p70 though on this climb 2 axes are often useful.
First ascent R.W. Head with J. Grange, A. Orset and J.M. Perrot, 13 August 1863.

The comfortable Cosmiques Hut, ideally situated at the starting point, makes this a very attractive traverse with an approach with no objective danger and little expenditure of energy. The route is more technical and marginally longer than the Goûter route, but the stages are of similar length; two hours to the Shoulder on the Tacul, two hours to the Col du Mont Maudit, perhaps a little over two to reach 4810m. To ensure that everything is to your advantage, it is important to be fully acclimatised as the cable car takes you from 1000m to 3800m in no time at all. If you are not fully prepared, you will soon become exhausted and slow, thus making the climb far more serious. As always in the Alps, maintaining a steady pace is the key to safety. This apart, the route is magnificent, with a gorgeous exposure to the rising sun and a staggering 360° panorama.

From the hut, go down to the Col du Midi (3532m) and climb the north side of the Tacul to the Shoulder (about 4050m, crevasses sometimes troublesome – see p64). Ignore any tracks heading east and descend south-east to Col Maudit (4035m). Climb diagonally WSW across northern slopes of Mont Maudit (some sérac fall danger) to Col du Mont Maudit, 4345m (possible ice and bergschrund difficulties). Traverse the west flank of Maudit to reach the Col de la Brenva (4303m). A steep pull up Mur de la Côte brings the summit slopes into view and a steady climb past Petits Mulets, takes you to the highest point. *Descents: pp74, 75.*

The traverse across Mont Maudit's north flank.

Mont Blanc from the top of the Mur de la Côte.

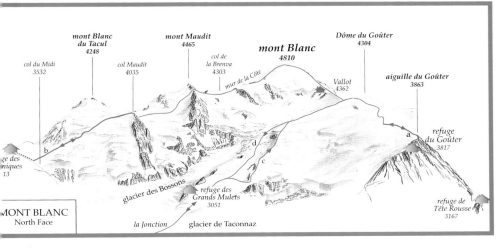

(c) Grands Mulets / Dôme du Goûter North Ridge / Bosses Ridge

Start   Grands Mulets Hut (3051m). Approach p95.

Difficulty   III/PD+ Snow ridge 40°–45°. Reaching the hut by La Jonction is rather chaotic, and the route's length, the effects of altitude and navigation problems in poor visibility are also factors. See p68.

Time   Hut to the Dôme, 4–5hrs, from there to Mont Blanc, 3hrs. Descent 4–5hrs.

Height gain   1000m to the Dôme, 750m for Mont Blanc (741m from Plan de l' Aiguille to the hut).

Conditions   A good frost (windslab risk after snowfall). Check La Jonction is passable. Acclimatisation essential.

Equipment   See Mont Blanc by the Goûter, p70.

First ascent   J. Balmat, J. Carrier, F. Paccard, J.M. Tournier, 8 June 1786 to the Col du Dôme. Bosses Ridge via Grands Mulets Glacier: E. Headland, G. Hodgkinson, C. Hudson, G. Joad with M. Anderegg, F. Couttet and two other guides, 29 July 1859. The described way by M.and Mme. Millot, M. Payot, H. Charlet, A. Payot, 19 August 1872.

The elegant North Ridge of the Dôme dominates the Chamonix side of Mont Blanc. Its airy trace, standing above the valleys and séracs, leads naturally to the Col du Dôme, then with total logic on to the Bosses Ridge and the coveted 4810 metres. The route is aesthetically pleasing with difficulties that are similar to those of the Normal Route on the Tacul – 45° slopes, fairly narrow ridge, simple line, wonderful and ever-changing views.

Turning back in case of bad weather is always possible and an escape to the Goûter Hut is an option once the Dôme is reached. Although sérac dangers can be avoided, those posed by windslab after snowfall should not be ignored. Good crampon work is also necessary. The height gain from the hut amounts to 1800m, so it is

important to be acclimatised, technically prepared, and well equipped with warm clothing.

It is worth emphasising the welcome and tranquillity to be found at the Grands Mulets Hut. It is easier to sleep at 3000m than at 3800m, and the approach walk has a unique atmosphere – a world of wild icy eccentricities where the crevasses at La Jonction can provoke a few cold sweats. Don't hesitate to call the warden who will be able to give you precise information on conditions. As a historical note, remember that the North Ridge of the Dôme was first climbed two months before the first ascent of Mont Blanc by Jacques Balmat with some rather unhelpful companions! So this regal old route offers historical interest as well.

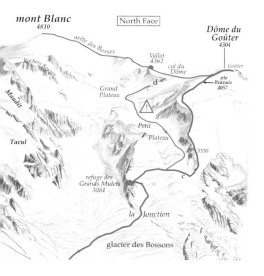

mont Blanc
4810

North Face

arête des Bosses

Dôme du
Goûter
4304

Rochers Rouges

Vallot
4362

col du
Dôme

Goûter

Maudit

Grand
Plateau

d

pte
Bravais
4057

Tacul

Petit

Plateau

3330

refuge des
Grands Mulets
3051

c

la Jonction

glacier des Bossons

Above the hut, climb south then south-west. At around 3200m, traverse west going slightly downwards, so as to pass below a little line of séracs which bars access to Rognon 3330 (45mins). Climb the slope to the right of the Rognon, turning any crevasses on their right and on the left gain the crest of the ridge that leads to Pointe Bravais (4057m, 3hrs): wide 40° slopes at the start (a few sections at 45°), the ridge becoming narrower and very fine between 3600 and 3800m, to finish on a long whaleback. From the Pointe Bravais, either:

Escape to the Goûter and the Nid d'Aigle:
Continue up the whaleback to near the summit of the Dôme (south-south-west), where the Normal Route on the west side is joined (see descents). Or:

Climb to the summit of Mont Blanc:
Make a slightly rising traverse southwards across the east side of the Dôme du Goûter, to reach the Col du Dôme (4237m, 30mins). From there, passing the Vallot Hut, climb to the summit of Mont Blanc by the Bosses Ridge and the West Ridge, 2½hrs from Pointe Bravais, 7–8hrs from the Grands Mulets Hut.

The third sunlit spur is the North Ridge of the Dôme. ▶

Descents from Mont Blanc
The choice of descent route must be made according to snow conditions. Diagram p73.

(a) To the Nid d'Aigle by the Aiguille du Goûter III/PD, 2400m. 4–5hrs. (The Grand Couloir crossing is badly exposed to stonefalls, some triggered by higher parties).
From the summit, follow the ridge, narrow at first, heading west then veering north-west, to cross over the Bosses in a descent to the Vallot Hut (4362m, emergency shelter). Continue north-west past the Col du Dôme, keeping on the right (north) past the Dôme du Goûter and then (taking care not to be lured off down the easier-angled North Ridge) descend steeper ground WSW to a shoulder that marks the start of its North-West Ridge. Go down the long slopes, keeping left, to reach the gently-angled ridge of the Aiguille du Goûter (3863m) and the hut below (3817m). Below the building, climb down the rocky rib (mixed at the start of the season, cables) for about 500m to the point where the Grand Couloir on the right is crossed (cable, stonefall risk). Shortly afterwards, go down the Tête Rousse Glacier (hut below at 3167m) and reach the Nid d'Aigle station by a good path (north-west then south-west).
Train to St. Gervais or to Bellvue station for the cable car to Les Houches.

On the North Ridge, above the Grands Mulets Rognon, seen on the right.

(b) To the Aiguille du Midi by reversing the Tacul-Maudit Traverse

III/PD+, 1110m, 4hrs. The descent from the Col du Mont Maudit can be tricky if iced (50°) or if the bergschrund is too wide – take gear for setting up one or two abseils, 30m and 50m). Also seek information beforehand about the state of the descent by the Normal Route on the Tacul. Diagram p73.

From Mont Blanc , descend north-north-east to the top of the Petits Mulets, then down steep slopes of Mur de la Côte to the Col de la Brenva (4303m). Go across (north-west then north) the western slopes of Mont Maudit, to the Col du Mont Maudit (4345m). Here the steep upper slope of the North Face of Mont Maudit forms a tricky obstacle and often requires two abseils: the first of 30m from a stake or Abalakov, the second of 50m equipped with pitons. Descend the face diagonally rightward, then in the centre (sérac fall risk). From the bottom, traverse east then north-east to the level of Col Maudit, and then climb north-west to reach the Tacul Shoulder (around 4050m). From there, descend the Tacul's Normal Route: generally straight down north, then soon diagonally right (north-east) to cross a first bergschrund around 3920m. Carry on north-east, to near the séracs bounding the face, so as to descend a steep cwm for 200m (north) to the big crevasses on the lower third of the face. Follow a long slope, steep to start with, which leads down (north-north-east) to the Col du Midi (3532m). Pass below the Cosmiques Hut, continue below the South-East Face of the Aiguille du Midi and then up, still north-north-east, to the North-East Ridge of the Aiguille du Midi, which leads back up to the cable car.

(c) To the Plan de l'Aiguille, by the Grands Mulets III/PD, 2500m of descent, 4–5hrs. A pleasant route at the start of the season when the crevasses at La Jonction are well covered. Ask at the Office de Haute Montagne or phone the Grands Mulets Hut.

From the summit, head west then north-west on the narrow then wider ridge, passing over the Bosses, then the Vallot Hut (4362m, emergency shelter and radio) and the Col du Dôme (4237m). From there, go down due east by the long slopes leading to the Grand Plateau. Continue due north down a series of slopes and easements, among them the Petit Plateau (around 3650m) where, for 150m–200m, the track is threatened by sérac falls from the left. After trending left for a section, move right to join the Grands Mulets rocks and find the hut (3051m). Then, making an S-turn from left to right, embark on the complex crevassed area of La Jonction. Cross this carefully heading north-east, to reach the Plan de l'Aiguille path (marked) just before the old Gâre des Glaciers (2414m, 3–4hrs from the summit). Reach Plan de l'Aiguille, in about an hour from where the cable car goes down to Chamonix. Diagrams pp73, 75.

Start Gonella Hut (3071m), 5hrs from the Cantine de Visaille, Val Veni, Italy. Approach see p95.
Difficulty III/PD on a heavily crevassed glacier. A long route at high altitude. See text on p68.
Time 6–8hrs. Descent to Aiguille du Midi 3–5hrs.
Height gain 3120m! 1370m to the hut then 1750m from hut to summit.
Conditions Stable snow, crevasses adequately covered (hazardous during an afternoon descent).
Equipment See Mont Blanc by the Goûter Route, p70.
First ascent First climbed in descent (known as Route du Pape because of the presence of Achille Ratti (later Pope Pius XI) with J. and L. Bonin, J. Gadin and A. Proment. 1 August 1890.

If you love wild spaces and tranquillity, this route is for you! The Route du Pape, first done by Achille Ratti, the future Pius XI, is the normal route for Italian alpinists.
At the end of Val Veni, on the other side of the Mont Blanc tunnel, the Miage face is a must to explore, although getting back to your car requires a bit of forethought.
You have to negotiate the labyrinth of the Dôme Glacier and come through the night to greet the dawn on the delicate Bionnassay Ridge to savour the final ridges leading up to the roof of Europe from a different perspective.

Route From the Gonella Hut, traverse north to reach the Dôme Glacier (15mins, inspect the day before). Climb the glacier on the right (west) bank, then in the middle, passing between the foot of the Tour des Aiguilles Grises (on the left), and a rocky rognon at the bottom of a buttress of the Dôme du Goûter (right). On a fairly steep slope, go left (west) to reach the Col des Aiguilles Grises (3810m), then follow the ridge north to the Piton des Italiens (4002m). Now ascend the Bionnassay/Dôme Ridge heading east to the Dôme du Goûter. Pass the Dôme's summit on the right to reach the Col du Dôme (4237m). Above this a short steep slope leads up to the Vallot Hut (emergency shelter). From there, go up the fine Bosses Ridge and then the narrow West Ridge to the summit of Mont Blanc. *Descents: pp74, 75.*

The south-west side of Mont Blanc showing the sunlit glacier (centre) taken by the Aiguilles Grises route.

Looking towards Dôme du Goûter from the Piton des Italiens.

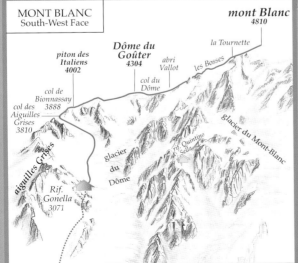

MONT BLANC
South-West Face

mont Blanc
4810

la Tournette

Dôme du Goûter
4304

abri Vallot

les Bosses

piton des Italiens
4002

col du Dôme

col de Bionnassay
3888

col des Aiguilles Grises
3810

glacier du Mont-Blanc

Tif Quintino Sella

aiguilles Grises

glacier du Dôme

Rif. Gonella
3071

SOUTH RIDGE
AND TRAVERSE

Start   The Durier Hut (3370m) which has a complex approach from the Val Montjoie, 7hrs (see p94) or from the Dôme de Miage.

Difficulty   IV/AD mixed, rock step 4a. The long hut walk, altitude, remoteness, corniced ridges, several steep slopes, make this a serious climb, particularly in deteriorating weather.

Time   From hut to summit 3–4hrs. Descent of East Ridge 1hr according to conditions. Then to the Dôme du Goûter allow 2hrs. From there Mont Blanc can be reached in 2hrs (4–5hrs for descent). If descending directly from the Dôme via the Goûter Hut to the Nid d'Aigle station (3hrs).

Height gain   Hut to summit of Bionnassay 680m, or to Mont Blanc 1550m. Hut approach 1400m.

Conditions   As for all high-altitude routes, the balance between sufficient snow cover, dry rock and ridges not too corniced or icy, requires good information, observation and judgement.

Equipment   Ice axes, rope, 3 ice pegs, slings, 4 quick draws, 4 karabiners, nuts, warm clothing..

First ascent   G. Gruber with K. Maurer and A. Juan, 13/14 August 1888.

You only need to spread out the map of the Mont Blanc range on the ground to see that its orographic layout enables all sorts of combinations. Linking together faces and glacier basins is possible, as Patrick Gabarrou did in 1988: seven north faces in ten days, including some of the most difficult. Or again, François Damilano in seven days in summer 1992: the complete traverse of the range by its ridges, from the Grands Montets to les Contamines, 40 summits and 13,000m of ascent and descent.

Less ambitiously, any climber in good training can look forward to two or three fine days linking up the ridges.

Aiguille de Bionnassay (left), Dôme du Goûter, Mont Blanc (right) and Dôme du Miages (foreground).

The ridge at the foot of the rock step.

The Durier Hut on the Col de Miage.

AIGUILLE DE BIONNASSAY
south-west aspect

Dôme
du Goûter
4304

aiguille
de Bionnassay
4052

refuge
du
Goûter

mont Blanc

piton des
Italiens
4002

col
3888

col des
Aiguilles Grises

glacier de Bionnassay italien

refuge
Durier
3370

col de
Miage

Aiguille de Bionnassay is the elegant and intriguing western satellite peak of Mont Blanc, well known for its North-West Face that is so well seen from the Goûter Hut and its approaches. Its South Ridge is the best ascent route and just to reach it (and the strategic Durier Hut) is an adventure in itself, whether you are coming from the Dômes de Miage or from Val Montjoie (p94). To spend the night at this eagle's nest is well worth the trip alone and ascending the South Ridge the following morning seems a generous bonus. The ridge, interesting in itself, is enhanced by views across the southern precipices of Mont Blanc and panoramic views of the western end of the range. The crux is the rock step, and this will be harder if snow-covered.

From the summit the impressively sharp and icy North-East Ridge leads down to the Col de Bionnassay and from there a fabulous snow crest rises steadily up to the Dôme du Goûter. A range of tempting options are then available – continuing to Mont Blanc and possibly on the Midi, exploring the intricacies of the Grands Mulets, or taking a rapid and convenient descent to the Goûter Hut. What is guaranteed on this magnificent traverse is a fine feeling of exposure – a magical interplay of altitude, space and light.

The view east from the Miage glacier basin over the Col de Mi

Route   From the Durier Hut, go up the South Ridge, snow or mixed, passing two shoulders. Follow a narrower ridge (cornices), then a short rocky rib. At the top of this, a horizontal snow ridge leads to the foot of the rock buttress (3780m, 2hrs). There are two possible routes:

(a) If snow conditions permit, it is quicker to turn it by a 150m traverse to the right (east). This is quite steep and exposed, but it leads to a series of snow couloirs (40°) which head up left to the top of the rock step.

(b) Start up a small dièdre blocked at the top by a jammed block (3b). Belay on a terrace. A flake system (3c/4a) is followed to the foot of a dièdre which is climbed to the top of the step (1hr, tricky, loose rock, some verglas.

A fine snaking snow ridge and then a steep slope lead to the very airy summit. 3–4hrs. from the hut.

Traverse and descent   Follow the very narrow and often corniced North-East Ridge, reach the Col de Bionnassay and go up to the Piton des Italiens on the east (mixed; 4002m). Here a descent of the Grises Route (p76), to the Gonella Hut and Italy is an option). Continue by the ridge which again turns north-east to a small snow saddle around 4200m, below the summit of the Dôme du Goûter (2hrs).

Mont Blanc, with the Aiguille de Bionnassay and its South Ridge (left) and Dome de Miage (right).

From the Dôme there are several options:

(a) *Continue to Mont Blanc*
Go up a little then traverse the slopes of the south side of the Dôme to reach the Col du Dôme. From there go up Mont Blanc by the Bosses Ridge (2hrs to the summit, descent 4–5hrs. See pp74, 75.

(b) *Descend to the Goûter Hut* and on to the Nid d'Aigle station, go north at the 4200m. contour, traversing the west side of the Dôme du Goûter. Near the rognon at 4140m, go west (turning an area of large crevasses)

and by the steep slope, descend left to reach the easy North Ridge of the Aiguille du Goûter and the hut situated just below (3817m, 40mins). Below the hut, go down the rocky rib (mixed at the start of the season, cables) for about 500m to the point where the Grand Couloir on the right is crossed (cable, stonefall). Shortly afterwards, go down the Tête Rousse Glacier (hut at 3167m on left bank) to the Nid d'Aigle station 2372m by a good path on the right bank – north-west then south-west. 2hrs from the Goûter Hut.

TRAVERSE

Start   Conscrits Hut (2602m); 4½hrs from Le Cugnon (of Contamines-Monjoie), see p94.
Difficulty   II/PD   A long exposed snow ridge sometimes corniced, and with a steep (45°) descent to the Col de la Bérangère.
Time   For the classic traverse, hut to hut, 6–8hrs.
Height gain   1090m (plus 1573m for the hut approach).
Conditions   June to September, easier when the glacier is more snowed-up.
Equipment   Ice axe, crampons, rope, 2 ice pegs, 2 slings, 3 karabiners.
First ascent   E.T. Coleman with F. Mollard and J. Jacquement, 2 September 1858.

Traversing long ridges always produces a liberated feeling, for without having to overcome any great technical difficulties, you have the sensation of travelling between heaven and earth. So the uncluttered ridges of the Dômes de Miage provide unforgettable inspiration to all those who love the beauty of nature. To reach them, you have to visit the Val Montjoie, pass through Les Contamines, and go up the majestic basin of the Glacier de Tré-la-Tête.

Once the heights are gained the traverse itself has no section of any particular difficulty, but that is all the more reason to be cautious with cornices and to move steadily and carefully. At the heart of the massif, the blue world of the séracs and the ice chutes may have unsettled you. Here, pausing to take stock and accustom yourself to the exposed ridges, let your gaze drift between two worlds, those of the valley and of the summits. The route may be done from east to west or vice versa. The former is more progressive as the harder sections are taken on the ascent

The start of the descent to the Col de la Bérangère.

The cwm below the Col des Dômes.

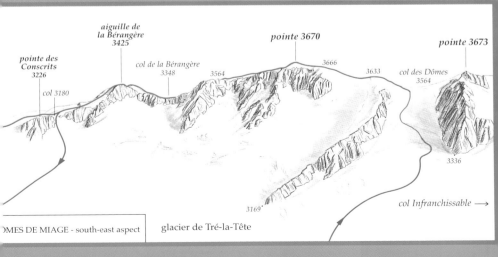

pointe des
Conscrits
3226

col 3180

aiguille de
la Bérangère
3425

col de la Bérangère
3348

3564

pointe 3670

3666

3633

pointe 3673

col des Dômes
3564

3336

3169

col Infranchissable →

ÔMES DE MIAGE - south-east aspect

glacier de Tré-la-Tête

and the descent to the hut is quick. From west to east is more technical on the descents, but goes more quickly to high altitude on the ascent to the ridge, with its view of Mont Blanc. The sole disadvantage is the long way back by the Tré-la-Tête Glacier when the snow will be slushy. On the other hand, this option becomes ideal if you are continuing to the Durier Hut to do the Aiguille de Bionnassay the next day. But that is a far bigger undertaking, requiring stable weather and a higher level of fitness and technique.

Route From the Conscrits Hut, head north across rocky outcrops and then north-east by an intermittent path (snow patches) to reach the Tré-la-Tête Glacier at around 2900m (1hr). Go up it on the right bank then, at around 3000m, go to the middle. Avoid an area of large crevasses by going right and at an easement (3300m) before the Col Infranchissable, branch off north-west to follow, first by the left bank and then in the middle, the fine valley leading to the Col des

Dômes (3564m). From here you can divert to the north-east and climb Dôme de Miage 3673m, the highest point (1hr there and back). Reach the Miages ridge a little south-west of the col (3½hrs from the hut). Follow the crest of the ridge crossing the magnificent Dôme Central (3633m), Pt. 3666 and Dôme Occidental (3670m) 4½hrs from the hut. Go down easily south-west, then by quite a steep slope (45°) reach the Col de la Bérangère. Turn a shoulder by easy rocks on the north side to reach the Aiguille de la Bérangère (1½hrs).

Descent From the summit of the Bérangère descend rocky zig-zags south to reach a snow slope (40° for 200m). From the bottom, go south-west down the Bérangère Glacier, then descend south down screes and rock to reach the hut. 45 mins.

The North Face of Miages and la Bérangère.

Starting up the ridge from the Col des Dômes.

LEX BLANCHE (3697m) AND DÔME DE NEIGE (3592m) TRAVERSE

Start Conscrits Hut (2602m); 4½hrs from Val Montjoie (hamlet of le Cugnon), p94.
Difficulty II/AD A long traverse mainly on snow with great technical variety, steep slopes both in ascent and descent (45° gully for 200m). The ridge has mixed ground and is sometimes exposed and corniced – seek current information.
Time From the hut to the top of the Aiguille Nord – 4½hrs. From there to the Dôme de Neige 3hrs, then 2hrs to descend the Tré-la-Tête Glacier.
Height gain Ascent 1300m. Descent 1400m to the glacier.
Conditions Stable snow, overnight freezing, and good weather forecast.
Equipment Crampons, ice axe, rope, 2 ice pegs, slings, 2 quick draws, 4 karabiners,
First ascent The north-west slope of Aiguille Nord, A.W. Moore, H. Walker with J. Anderegg and J. Jaun, 23 July 1870.

The Tré-la-Tête to Lex Blanche section of the traverse seen from the Conscrits Hut.

Finding a different and little-known classic in the Mont Blanc massif is not easy. The Tré-la-Tête Nord – Lex Blanche – Dôme de Neige traverse is remote but has many attractions. Now (as a result of this guidebook attention) it may soon attract crowds – the eternal problem with guidebooks.

The climb takes a section of the frontier ridge that borders the Tré-la-Tête basin on the east. At present you will have a good chance of being alone along this 3km airy crest and perhaps even have to break the trail, do some route-finding, inspect map and guide-book – demands that are almost unknown on many similar

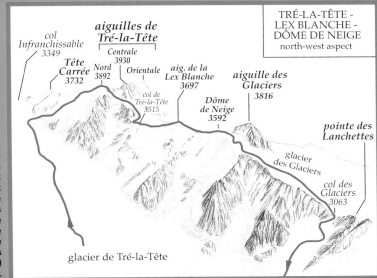

**TRÉ-LA-TÊTE -
LEX BLANCHE -
DÔME DE NEIGE**
north-west aspect

col
Infranchissable
3349

*aiguilles de
Tré-la-Tête*

Centrale
3930

Tête
Carrée
3732

Nord
3892

Orientale

*aig. de la
Lex Blanche*
3697

*aiguille des
Glaciers*
3816

col de
Tré-la-Tête
3515

Dôme
de Neige
3592

*pointe des
Lanchettes*

glacier
des Glaciers

col des
Glaciers
3063

glacier de Tré-la-Tête

The southern section of the traverse. The Aiguille de la Lex Blanche is on the left and in the centre the rocky Aiguille des Glaciers peeps above the Dôme de Neige descent ridge. The Col des Glaciers basin (of the descent) is on the far right.

routes, where you just follow the tracks. Though remote and still wild in concept, confidence is instilled by the reassuring presence of the hut below, and the moderate difficulties (a full range of problems on mainly snow and mixed ground, gullies and a hanging balcony). Nevertheless there remains a certain caution as the scene, though magnificent, remains somewhat intimidating and there is always a nagging doubt on a route where retreat is difficult. The satisfaction of achieving three summits in succession after a long day at altitude amply repays this commitment. So with good weather, the magnitude of the undertaking should not deter, and a fine adventure awaits you.

Route

From the hut go north-east by outcrops and an intermittent path (snow patches) to reach the Tré-la-Tête Glacier around 2900m (1½hrs). Go up the right bank at first, then around 3000m move to the middle. Turn an area of large crevasses on their right and at around 3300m reach the easement before the Col Infranchissable. 2½hrs from the hut.

Aiguille Nord de Tré-la-Tête by its north-west face Start 100m right of the west spur of Tête Carrée, in the middle of the face. Go diagonally right to turn a first sérac barrier and above this come back left (50° slope). Cross a bergschrund left of the second

sérac wall and go up slightly rightwards. After crossing a second bergschrund, climb a short steep slope to the snowy summit ridge. Follow it to the summit (3892m 2hrs), 4½hrs from the hut.

Traverse From the summit, go south-east down the corniced ridge 50m heading for the Aiguille Central and descend southwards in a steep snowy couloir leading to a hanging balcony. Contour across it south-south-east and then follow the leftmost of the two snow slopes (steep for about 150m) to reach the upper part of the Lex Blanche Glacier. Go up it to the west to the Col de Tré-la-Tête (3515m). Go back on to the summit ridge which is followed south. After some snow, turn the rocks by the west side to follow the ridge, which again becomes snowy and quite steep, to the summit of the Lex Blanche, 3697m – named Lée Blanche on the IGN map. Descend the ridge (cornices), reach the Col de la Scie and continue on the snow ridge south then south-west to the Dôme de Neige (3592m, 3hrs) bypassing the summit buttresses of the Aiguille des Glaciers 3816m to the east.

Descent Do not follow the track on the Glacier des Glaciers (south) but keep as far as possible to the snow ridge coming down south-west from the Dôme (a wide, easy, slope to begin with). It passes near the Col du Moyen Âge (gap) and narrows to arrive, after an exposed section (poor rock), at Col des Glaciers (3063m). 1hr.
Descend the slope on the north side, steep to start with, keep left for 300m, and then come back right to reach the Tré-la-Tête Glacier. Cross it and on the right bank cross the Tré-la Grande fault (1hr, 2hrs from the Dôme de Neige). From there, either descend to Le Cugnon or go back up to the Conscrits Hut.

The crest of the Conscrits Hut.

The ridge leading up to the Dôme de Neige des Glaciers. Aiguille des Glaciers is the higher rock peak beyond.

NORTH-WEST FACE

Start Conscrits Hut (2602m), 4½hrs from Val Montjoie. See p94.
Difficulty II/AD- to AD depending on snow/ice conditions, particularly on the final section below the summit (400m at 50°). Two or three bergschrunds to cross. Quite a long descent with a ridge traverse and some steep slopes below the Col des Glaciers.
Time Approach 1½hrs. Ascent 3hrs. Descent 3hrs to the Tré-la-Tête Hut.
Height gain 800m from bergschrund to summit.
Conditions Best in early season, when snow covers bergschrunds and the central rock belt.
Equipment Axes, rope, 6 ice pegs, selection of slings, 6 quick draws, 5 karabiners.
First ascent R. Gaché, P. Gayet-Tancrède and R. Jonquière, 31 July 1931.

Seen face-on, snow or ice faces are always impressive. Festooned with séracs and laced with debris chutes, they look dangerous and scary seeming to epitomise the void with their smooth and often shining icy visage — like huge mirrors. How can a tiny human being dare to go there? With a bit of experience, you soon learn to judge the real angles, and with two or three classics under your belt this optical illusion will no longer trouble you.

The North Face of Lex* Blanche is a typical case. Its slopes steepen progressively from the bottom to the sharp summit. A few features stand out to break the monotony that a uniform slope might have created. The rock spurs guide you, the rock band is a landmark and the summit slope is the icing on the cake.

Situated in the middle of the long barrier that bounds the Tré-la-Tête Glacier on the south, this elegant North Face possesses panache and character, requiring from you good technical ability — members of the party should be able to move together or take belays on ice axe or ice pegs, maintaining a steady progress so that the 800m face is climbed efficiently.

The outing does not finish at the summit, as there is an easy but airy traverse to the Col des Glaciers, for the return to Tré-la-Grande. After coming out into the sun, a vast panorama is revealed, where your gaze will lose itself in the Mediterranean mists.

*Lée on IGN.

Route From the hut, go up north then east by rock outcrops and a path to reach the Tré-la-Tête Glacier around 2750m (1hr). Cross it (east) to reach the bottom of the face. Start up the centre of the glacier bay between the West Ridge of the Lex Blanche and a spur coming from the lower first third of the face (35°– 40°, sometimes 2 bergschrunds). On reaching the rock band that closes the bay, traverse left by bands of snow and come out on top of the spur. Follow the vague ridge which sticks out of the face, alongside some fine séracs, then go straight up the slope which gets steeper (50° for 400m, with sometimes a bergschrund) to finish at the summit. 4hrs.

Descent Go south along the corniced ridge to the Dôme de Neige (30mins). Descend the slopes leading south-west, keeping close to the crest of the ridge, which narrows after the Col du Moyen Âge and leads, after an exposed section on poor rock, to the Col des Glaciers (3063m, 1hr).

Descend the slope on the north side, steep at the start, bear left for 300m and come back right to reach the Tré-la-Tête Glacier. Cross it, and on the right bank, cross the fault of Tré-la-Grande then back to the hut. 3–4hrs from the summit of the Lex Blanche.

La pente de sortie.

**AIGUILLE DE LA
LEX BLANCHE**
North-West Face

*aiguille
de la
Lex Blanche*
3697

Dôme
de Neige

2908

2880

glacier de Tré-la-Tête

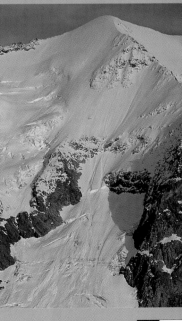

NORMAL ROUTE

Start Tré-la-Tête Hotel (1970m); reached in 2hrs from Le Cugnon (to the south of Les Contamines-Montjoie). See p95.
Difficulty II/F Glacier approach, snow slopes at 35–40°, rocky summit ridge, a little exposed but easy.
Time From hut to summit 4–5hrs. Descent 2–3hrs.
Height gain 1226m.
Conditions Best in early season when there is stable snow and before the slopes become too dry (ice, stonefall).
Equipment Ice axe, rope, 2 ice pegs, 2 slings, 4 karabiners, helmet, headlamp.
First ascent A. Hutter and Captain X with A. Magnin and N. Allantaz, 19 August 1894.

Mont Tondu offers the advantage of being a wonderful viewpoint over an ancient landscape, while presenting only moderate technical difficulties. It is the ideal training peak. The route is quite obvious, which enables you to concentrate on technical considerations: maintaining rhythm in cramponing, synchronising this with the movement of the ice axe and regular breathing, the alternation of rope handling, either run out or taken up in coils, depending on terrain.

A useful procedure on the evening prior to a climb, is to make a reconnaissance of the way to the glacier and thereby remove uncertainties, enable you to sleep untroubled, so that in the small hours, under starlight, you will be moving over familiar terrain. At the cairn on the Pain de Sucre, don't stay there as many do, but follow the easy rock ridge to the summit, a bit off centre to the south. The view from there to Mont Blanc is very interesting as it reveals the profiles of both the French and Italian Normal Routes. If one makes an early start from the valley the ascent of the Tondu can easily be linked (as a training climb) to an ascent to the Conscrits Hut.

On the return, once at the foot of the narrows, an hour is then enough to reach the hut, with a route in the bag, before embarking on the lovely traverse of the Dômes de Miage the following day.

Another option is to use the Col Mont Tondu to reach the Robert Blanc Hut from where the Aiguille des Glaciers could be climbed

next day followed by a descent to the Tré-la-Tête Glacier by the Col des Glaciers, thereby completing a clever round.

Route From the hut follow the path to the east along an old moraine. After a few hairpins the path crosses some escarpments that stand above the glacier ravine (uphills, descents, markers and cairns). Reach the glacier (30mins from the hut). Go up the glacier, preferably in the middle, then go left at around 2300m to climb the sérac barrier of Tré-la-Grande near the rocks on the right bank. Then go south-east to start the climb proper at around 2500m at the narrows to the right of the North Spur of Aiguille des Lanchettes). 1½hrs.

Ascent Go up the narrows diagonally right and get on to the Mont Tondu Glacier, which is climbed by the centre to the col (around 3000m) between the Pain de Sucre and the Pyramide Chaplan. Go south on this easy snow ridge to the big cairn on the Pain de Sucre (3169m). Finally, follow the rock ridge to the summit (mainly on the right side, easy, sometimes exposed). 4–5hrs from the hut.

Descent By the same route. 2–3hrs.

Variation The NW Ridge gives a mixed climb that is a little more technical (rock step). Same approach as the Normal Route.

The Conscrits Hut with the north-east side of Mont Tondu in the background.

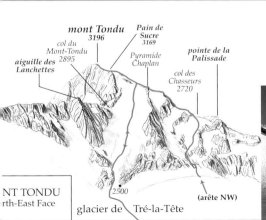

mont Tondu
3196

Pain de
Sucre
3169

col du
Mont-Tondu
2895

aiguille des
Lanchettes

Pyramide
Chaplan

pointe de la
Palissade

col des
Chasseurs
2720

NT TONDU
rth-East Face

2500

(arête NW)

glacier de Tré-la-Tête

MAPS, GPS

The general line of routes is described in the text, but this should be supplemented by the 1:25,000 IGN maps: *Chamonix* 3630 OT and *Saint-Gervais* 3531 ET and also GPS points: UTM format, WGS 84 system. GPS accuracy is + or −15m and variations in interpretation by users may result in a margin of error of up to 40m. Remember this only fixes position, not height so compass and altimeter remain useful instruments.

HUT APPROACHES AND BOOKING DETAILS

(*Abbreviations:* CAF, CAI, CAS for the French, Italian, and Swiss Alpine Clubs. *Book in advance during busy periods.*)

Albert Premier Hut (Refuge Albert 1er) 2702m

Tel. (00)33 (0)4 50 54 06 20. Guardian: summer and some spring weekends. CAF 150 places.
Start: At the village of Le Tour (20km north of Chamonix).
Route: Reach the Col de Balme by the uplift Charamillon-Balme (gondola then chairlift). Then follow a good path south-east then south, on a rising traverse. Around 2330m it goes round the Bec de Picheu and heads east, then crosses an area of screes (or snow) to climb the moraine below the hut. 1½–4hrs, depending on whether lifts are used. *Note:* In winter/spring, access on skis is from the Col du Passon starting from the Grands Montets.
GPS 32T E 0344 117 N 5095 662

Argentière Hut (Refuge d'Argentiere) 2771m

Tel. (00)33 (0)4 50 53 16 92. Guardian in summer and February – May. CAF 140 places. Glacier travel.
Start: Argentière for the Grands Montets cable car.
Route: From the upper station, go down to the Col des Grands Montets (3233m). Cross the bergschrund and go north-east down the Glacier des Rognons to the top of the moraine (2754m). Go down its crest, and reach the Argentière Glacier. Climb south-east up the left bank and around 2700m go east to the moraine on the right bank and the hut 70m above this. 1½hrs.
GPS 32T E 0345 345 N 5090 015

Conscrits Hut (Refuge des Conscrits) 2002m

Tel. (00)33 (0)4 79 89 09 03. Guardian: April to end of September. CAF 90 places. Glacier approach.
Start / Route: The same as for Tré-la-Tête Hut (2hrs). From there, go eastwards up the path which, after a few hairpins, crosses broken ground above the Tré-la Tête gorge. Go onto the glacier and up the middle and then, below the Tré-la-Grande fault, go diagonally left to reach rocks on the right (north) bank. Go up these by a ramp on the left (moraine, chains). Follow a path eastwards, which then turns up to the north-east over rocky outcrops. The hut is just after the rain gauge (2½hrs). 4½hrs from Le Cugnon. *Note:* In winter take glacier's right bank after the Tré-la-Grande séracs, to climb one of the couloirs south of the hut.
GPS 32T E 0326 302 N 5072 800

Cosmiques Hut (Refuge des Cosmiques) 3613m

Tel. (00)33 (0)4 50 54 40 16. Guardian: February – October. Private, 148 places [No concession for reciprocal rights cards]. Snow ridge/glacier approach.
Start: Chamonix-Sud, Aiguille du Midi cable car.
Route: From the Aiguille du Midi (3795m) descend the North-East Ridge (exposed, care required) to the first flattening. Turn down the slope to the right, south then south-west. Pass below the Midi's South Face and by a slight re-ascent, reach the hut. 20mins.
GPS 32T E 0335 896 N 5082 115

Couvercle Hut (Refuge du Couvercle) 2687m

Tel. (00)33 (0)4 50 53 16 94. Guardian: Throughout the summer, and weekends in May. CAF 120 places. Glacier travel and rock step (steep ladders).
Start: Chamonix, Montenvers/Mer de Glace station.
Route: From Montenvers (1913m), go south to ladders leading down to the Mer de Glace. Go up it by the left bank, then move to the middle, to gradually go on to the right bank at around 2050m. Go south-east across moraines, keeping left to meet the bottom of the Egralets rock step (2230m). Climb this by steep fixed ladders, go north-east up terraces, then by the path to the hut. 3½hrs. In winter/spring, access on skis is by the moraine on the left bank of the Talèfre Glacier.
GPS 32T E 0342 205 N 5086 095

Durier Hut (Refuge Durier) 3370m

Tel. (00)33 (0)6 89 53 25 10. Guardian in summer. CAF 17 places. Long approach with steep loose rock.
Start: Gorges de la Gruvaz car park (Val Montjoie).
Route: Follow the trail south, then north-east, to the Miage Chalets (1hr). Go along the left bank of the river (bridge) and continue up the right bank, to the start of a little path to the right of a waterfall. Follow it and climb the moraine, due east. Around 2670m, cross the snow below the bottom of the Tricot Glacier to reach the moraine on the opposite side (big cairn, col). Go up this moraine until just below the Plan-Glacier Hut, which is 50m higher up to the east (3hrs). Descend to reach the Miage Glacier at the Plan Glacier plateau around 2550m. Go up it (south-east, falling ice), to the bottom of the rocky central spur (around 2800m, paint marks). Climb the spur (poor rock), keeping left higher up on steeper ground, before coming back right to reach a snow saddle and the hut (3hrs). 7hrs. Note: It is also possible to come via the Miage Glacier. From there, a marked and fixed route leads at around 2800m to the glacier, which is contoured to the bottom of the central spur. Longer and more technical.
GPS 32T E 0329 975 N 5076 942

Gonella Hut (Rifugio Gonella) 3071m

Tel. (00)39 0165 885 101 Also called the Dôme Hut. Guardian in summer. CAI 70 places. Glacier travel.
Start: Cantine de la Visaille 1670m. (Val Veni, Italy).
Route: From the car park go up the road to the Lac de Combal. Before the bridge which crosses the Doire, take the trail on the right which goes up to the Lac de

Miage. By the path behind the refreshment hut, get on to the crest of the moraine and reach the Miage Glacier (boulders). Go up the glacier to around 2600m. A track can be seen going up right, on the west side of the Aiguilles Grises ridge. Follow it (some fixed gear), then go up a snow slope on the right bank of the Dôme Glacier, to climb the rock spur to the hut. 5hrs.
GPS 32T E 0331 600 N 5076 285

Goûter Hut (Refuge du Goûter) **3817m**
Tel. (00)33 (0)4 50 54 40 93. Guardian: June – October. CAF 116 places. Hazardous mixed ground.
Start: Les Houches, Chamonix Valley (Telepherique to La Chatlette to join the Tramway) or the Tramway alone from Le Fayet or St-Gervais (Val Montjoie).
Route: From the Nid d'Aigle, terminus of the Tramway du Mont Blanc (2372m), go south-east then north-east on the slopes bordering the Rognes ridge. After the forestry hut (2768m), traverse south-east, then go up some rocks on the edge of the Griaz Glacier, to around 3140m. 2hrs (Tête Rousse Hut on the opposite side). Go up the glacier diagonally rightwards to reach the right bank of the big couloir coming down from the Aiguille du Goûter. Cross the couloir (stonefall danger, cable), and climb the wide rib on its left bank (loose rock, cables) to the hut. 5hrs.
GPS 32T E 0331 530 N 5080 063

Grands Mulets Hut (Ref. des Gr. Mulets) **3051m**
Tel. (00)33 (0)4 50 53 16 98. Guardian: April – mid-September, depending on the state of the Glacier. Glacier travel, sometimes complex.
Start: Plan de l' Aiguille, middle station of the Aiguille du Midi cable car.
Route: Follow the path which goes up south-eastwards and leave it around 2370m (15mins), to go south-west across the moraine (2385m). Go down on to the Pèlerins Glacier and cross it. After crossing the moraine on its left bank, follow the path that passes the old Gare des Glaciers, then leads to the Bossons Glacier, around 2500m. Cross it (south-west) and at about 2650m head south to cross the very crevassed area of la Jonction. Then head SSE for the Grand Mulet rock to reach the far south side of the rognon. Go up slanting leftwards (fixed gear) to the hut. 3hrs.
GPS 32T E 0333 984 N 5081 465

Leschaux Hut (Refuge de Leschaux) **2431m**
Tel. (00)33 (0)6 73 10 29 47 or (0)6 99 59 71 67. Guardian in summer. CAF 10 places. Glacier travel.
Start / Route: Follow the Montenverts / Mer de Glace approach for the Couvercle Hut to the moraine on the right bank below the Egralets (1½hrs). Continue south-east, to go first up a subsidiary glacier tongue, and then the Leschaux Glacier itself. Bear steadily towards its right bank, to reach the bottom of a couloir (avalanche-prone) to the north of the hut. Go up the debris cone to reach the path on the right (ladder and gangway at the start), leading to the hut. 2½hrs from Montenvers.
GPS 32T E 0343 350 N 5084 356

Torino Hut (Rifugio Torino) **3371m**
Tel *(Summer Hut)* (00)39 0165 844 034. Guardian: June – September. CAI 160 places.
Tel *(Winter Hut)* (00)39 0165 846 484. Guardian: October, December – June. 40 places.
Start / Approach: The Funivie Monte Bianco from La Palud (Italian Val Ferret). From the Torino upper station, reach the winter hut on foot, or the summer hut by a stairway in a tunnel (10mins). From the Aiguille du Midi on the French side either approach on foot across the Vallée Blanche (2½hrs) or use the Mont-Blanc gondola.
GPS 32T E 0339 548 N 5078 930

Tré-la-Tête Hut (Ref./Hotel de Tré-la-Tête) **1970m**
Tel. (00)33 (0)4 50 47 01 68. Guardian in summer and some periods in the spring. Private, 80 places.
Start: Le Cugnon (in the south of Val Montjoie).
Route: Climb north-east on a good path, then by hairpins to the east before a long rising traverse eastwards leads to the hut. 2hrs.
GPS 32T E 0323 939 N 5073 414

Trient Hut (Cabane du Trient) **3170m**
Tel. (00)41 (0)27 783 14 38. Guardian spring and summer. CAS 150 places.
Start: Champex (Switzerland), La Breya chairlift.
Route: From the top station (2188m), go SW along the path to reach the centre of the Combe d'Orny, which is climbed westwards. Go past the lake and following the moraine, reach the Orny Hut (2hrs). From there, follow the left bank to the Col d'Orny (3098m). Then go diagonally north to the hut (easy rocks, 1hr), or go up right before the col, around 2860m, by a good path. 3hrs.
GPS 32T E 0348 517 N 5095 850

CABLE CARS / CHAIR LIFTS / MT. RAILWAYS
Opening is seasonal: check locally for dates and times.

Aiguille du Midi (00)33 (0)4 50 53 30 80
Chamonix-Sud, 2-section cable car. Connection to the Pointe Helbronner in Italy by gondola.

Bellevue (00)33 (0)4 50 54 40 32 Les Houches,

Charamillon-Balme (00)33 (0)4 50 54 40 32
Le Tour, gondola then chairlift.

Helbronner-Funivie Monte Bianco
(00)39 (0)165 89 925 La Palud in Italy, 3-section cable car. Connection to Aiguille du Midi by gondola.

La Breya (00)41 (0)27 783 13 44
Champex in Switzerland, chairlift.

Lognan-Grands Montets (00)33 (0)4 50 54 00 71
Argentière, 2 section cable car.

Montenvers - Mer de Glace (00)33 (0)4 50 53 12 54
Chamonix, rack railway

Nid d'Aigle-Tramway du Mont Blanc (TMB)
(00)33 (0)4 50 47 51 83 Le Fayet, rack railway. Connection with Les Houches by Bellevue cable car.

INFORMATION

Chamonix Tourist Office
(00)33 (0)4 50 53 00 24 www.chamonix.com

Les Houches Tourist Office
(00)33 (0)4 50 55 50 62 www.leshouches.com

Saint-Gervais Tourist Office
(00)33 (0)4 50 47 76 08 www.st-gervais.net

CAF Mont Blanc huts Can be consulted about mountain conditions (00)33 (0)4 50 53 16 03

OHM (Office de la Haute Montagne)
Place de l'Église, Chamonix: Gathers climbing information about current difficulties and conditions. (00)33 (0)4 50 53 22 08. www.ohm-chamonix.com

Guides' Syndicate
www.montagneinfo.net

WEATHER

Météo France Tel. summary 3250 or www.meteo.fr
Haute Savoie recorded message. (00)33 (0)892 68 02 74
Météo Val d'Aoste: www.regione.vda.it
Météo Alpes Romandes: (00)41 848 800 162
www.meteosuisse.ch

RESCUE

International Rescue numbers from a mobile phone France 112, Italy 118, Switzerland 144
PGHM (Peloton de Gendarmerie de Haute Montagne)
Chamonix: (00)33 (0)4 50 53 16 89
St-Gervais: (00)33 (0)4 50 78 10 81

BIBLIOGRAPHY (in English)

The Mont Blanc Massif / The Hundred Finest Routes
Gaston Rébuffat (Bâton Wicks, 2005). Timeless classic updated to cover climate changes – with rock routes and *grandes courses* as well as classic snow/mixed
Mountain Rescue, Chamonix·Mont Blanc
Anne Sauvy (Bâton Wicks, 2006). Accidents, rescues and incidents on many of the popular routes.
Conquistadors of the Useless Lionel Terray (Bâton Wicks/The Mountaineers, 2001).
Mont Blanc Massif / Selected Climbs 2 vols. Lindsay Griffin (Alpine Club, 1990, 1991). The main guides to the range in English with climbs of all grades.
The Alpine 4000m Peaks by the Classic Routes
Richard Goedeke (Bâton Wicks/Menasha Ridge 2006).
Snow Ice and Mixed: The Guide to the Mont Blanc Range 2 vols. François Damilano (J.M. Editions, 2006).

THE AUTHORS

Jean-Louis Laroche is mountain guide, photographer and free-lance journalist. As a Mont Blanc enthusiast for over thirty years he has climbed most of the classics on rock, ice, snow and mixed ground. He has pioneered new rock routes in the pre-Alps, as well as ice-falls and has been on expeditions to remote areas.
Florence LeLong is a free-lance photographer, illustrator, mountain leader and alpinist who seeks to share with us her passion for the realms of nature which light reveals to us. Co-author of several photographic books, she also exhibits and does press and publishing work. Laroche and LeLong run a business called *Alpinisme-Escalades*.

RISKS SAFETY

The routes in this book are described purely for information. The climber, thus informed of the risks, remains entirely responsible for his / her own safety and for the choices made according to personal abilities or the hazards inherent in the mountain conditions encountered. The routes are described in terms of a particular set of conditions at specific times and it is essential to ensure, before setting out, that the routes have not changed due to rockfall, removal of equipment or whether they may not be done for one or another reason. The authors and editors accept no responsibility in case of accident or incident on the routes described in this work. Given that this book is specific to the time of its publication, it may in no circumstances be used as a reference in legal cases. *Note for the UK/US edition:* It is assumed that all using this guide will be equipped with helmets, head torches, crampons, ropes and hardware and be proficient in their use. They should also have suitable clothing, snow glasses/goggles, sun cream, food/drink and emergency equipment and be aware of the need for acclimatisation. Those unfamiliar with alpinism might consider hiring a guide, attending a training course, or spending a day on easy slopes below the Aiguille du Midi or Grandes Montets téléphériques, the Torino Hut or other suitable places for practising and perfecting basic skills – cramponing, crevasse-belaying and rescue, ice-axe braking, moving together when roped and taking emergency belays – before attempting the described expeditions. Time spent on training (even for the experienced climber returning to alpinism after a break) is rarely wasted and, indeed, removes many of the fumblings and uncertainties of nightime starts and glacier approaches in this hazardous terrain. In this respect inspection of initial terrain leading from hut to glacier on the day before the climb is usually worthwhile. The choice of what permutation of ice axes and ski sticks to take on also needs consideration – short technical axes are less useful on the easier-angled terrain found on F and PD climbs (and descending steep paths and scree slopes) where a longer axe and one or two ski sticks may be better.

Printed in Spain in 2007 in association with Editions Desnivel